D0857558

ge Library

Chemical Fundamentals of Geology

Second edition

Robin Gill

Department of Geology, Royal Holloway, University of London, UK

CHAPMAN & HALL

London · Glasgow · Weinheim · New York · Tokyo · Melbourne · Madras

Published by Chapman & Hall, 2–6 Boundary Row, London SE1 8HN, UK

Chapman & Hall, 2–6 Boundary Row, London SE1 8HN, UK

Blackie Academic & Professional, Wester Cleddens Road, Bishopbriggs, Glasgow G64 2NZ, UK

Chapman & Hall GmbH, Pappelallee 3, 69469 Weinheim, Germany

Chapman & Hall USA, 115 Fifth Avenue, New York, NY 10003, USA

Chapman & Hall Japan, ITP-Japan, Kyowa Building, 3F, 2-2-1 Hirakawacho, Chiyoda-ku, Tokyo 102, Japan

Chapman & Hall Australia, 102 Dodds Street, South Melbourne Victoria 3205, Australia

Chapman & Hall India, R. Seshadri, 32 Second Main Road, CIT East, Madras 600 035, India

First edition 1989

Reprinted 1991

Second edition 1996

© 1989, 1996 Robin Gill

Typeset in plantin by Best-set Typesetter Ltd., Hong Kong

Printed in England by Clays Ltd, St Ives plc

ISBN 0 412 54930 1

Apart from any fair dealing for the purposes of research or private study, or criticism or review, as permitted under the UK Copyright Designs and Patents Act, 1988, this publication may not be reproduced, stored, or transmitted, in any form or by any means, without the prior permission in writing of the publishers, or in the case of reprographic reproduction only in accordance with the terms of the licences issued by the Copyright Licensing Agency in the UK, or in accordance with the terms of licences issued by the appropriate Reproduction Rights Organization outside the UK. Enquiries concerning reproduction outside the terms stated here should be sent to the publishers at the London address printed on this page.

The publisher makes no representation, express or implied, with regard to the accuracy of the information contained in this book and cannot accept any legal responsibility or liability for any errors or omissions that may be made.

A catalogue record for this book is available from the British Library

Library of Congress Catalog Card Number: 95-74644

∞ Printed on permanent acid-free text paper, manufactured in accordance with ANSI/NISO Z39.48-1992 and ANSI/NISO Z39.48-1984 (Permanence of Paper).

Scilib
QD
39.2
.G55
1996

070896-272867

Contents

iii

CARLETON COLLEGE LIBRARY

CONTENTS

Preface to the First Edition

Chemical principles are fundamental to a large area of geological science, and the student who reads for a degree in geology without knowing any basic chemistry is handicapped from the start. This book has been written with such students in mind, but I hope it will also provide a helpful refresher course and useful background reading in 'geo-relevant' chemistry for all earth science students who believe in understanding rather than merely memorizing.

The book has been conceived in three broad sections. The first, Chapters 1–4, deals with the basic physical chemistry of geological processes, emphasizing how consideration of energy can aid our understanding. The second section, Chapters 5–8, introduces the wave-mechanical view of the atom and describes from that perspective the various types of chemical bonding that give minerals their distinctive properties. The final section, Chapters 9 and 10, surveys the geologically relevant elements, concluding with a chapter on why some are more abundant than others, in the universe as a whole and in the Earth in particular. The emphasis throughout is on geological and environmental relevance; laboratory reactions are hardly mentioned.

The book is designed to be accessible and stimulating even to students who remember none of their school chemistry; for these readers, introductory background material and a glossary are provided in the Appendices. The more advanced material, which I hope will sustain the interest of the chemically literate reader, has been segregated into boxes that can be ignored on a first reading. The text is thoroughly cross-referenced to help the browser or the searcher to find what each wants. The book might also assist teachers of more advanced courses in geochemistry, mineralogy and petrology by relieving them of the need to cover elementary material in class.

Many colleagues have given me advice and encouragement during the writing of the book, and I am particularly grateful to the following colleagues who have read and commented on individual chapters and suggested innumerable improvements: David Alderton, Peter Barnard, Keith Cox, Giles Droop, Paul Henderson, Steve Killops, Robert Hutchison, Philip Lee and Eric Whittaker. Professor W. D. Carlson, Dr T. K. Halstead and Dr J. B. Wright read the whole manuscript and provided a wealth of helpful comment. The errors and eccentricities that remain are of course my responsibility alone. The book has grown from a lecture course I gave at the former Geology Department of Chelsea College, and I warmly acknowledge the opportunities which that department provided.

I would like to thank Joan Hirons, Sue Clay and Jennifer Callard for typing some of the chapters, and Neil Holloway and Christine Flood who prepared a number of the figures. I owe a great deal to Roger Jones of Unwin Hyman, whose faith in the project sustained me in the face of initial difficulties.

The biggest debt I owe is to Mary, Joanna and Tim, who have tolerated with remarkably good humour my neglect of family activities, the perpetual retreat up to the study, and the curtailment of holidays. I promise not to write another book for a long time.

Preface to the Second Edition

Since its first publication in 1989, this book has been used by a much wider range of students than I originally envisaged. My initial objective was simply to equip elementary Earth science undergraduates with the basic chemical understanding that their geological degree programme would require, but in the course of writing the first edition I fell into the trap of trying to entice them a little further so that they would embrace some of the underlying scientific principles. Accordingly, though somewhat to my surprise, the book has been adopted at the advanced undergraduate level and even as remedial reading for post-graduate studies as well as for its original purpose. It has also been adopted in some chemistry departments as a geochemistry primer.

In preparing this edition I have tried to increase the book's usefulness in two directions. Firstly, I have sought to make it more reader-friendly to those, such as mature students, who may remember very little chemistry from their school days or whose experiences of school chemistry were discouraging. Secondly, with more advanced students in mind, I have introduced a little more chemical rigour and more pointers to geological and geochemical applications, and have, I hope, also reflected some recent advances in cosmological, geological and environmental understanding. The change in format has allowed me to make more effective use of the boxes and to enhance some figures.

I am grateful to many people who have fed back to me their reactions to the first edition, particularly Paul Browning, Hilary Downes, Mike Henderson, Bob Major, Steven Richardson and Andy Saunders. Not every suggestion, of course, has seemed to me consistent with the original spirit of the book, but all comments have been valuable in giving me a reader's or a teacher's eye view. The encouragement I have received from colleagues such as Derek Blundell and Euan Nisbet has also been greatly appreciated. My family has been as understanding (or as resigned) as ever!

PREFACE TO THE SECOND EDITION

I thank Lynne Blything for her kind help with figure revision, and Ruth Cripwell and Ian Francis for their editorial advice and encouragement.

List of tables

List of boxes

1
Energy in geological processes

1.1 Introduction

The purpose of this book is to introduce the ordinary geology student to chemical principles which are fundamental to the science of geology. There can be no more fundamental place to begin than with the topic of energy, which lies at the heart of both geology and chemistry. Energy plays a role in every geological process, from the atom-by-atom growth of a mineral crystal to the elevation of entire mountain chains. Consideration of energy provides an incisive intellectual tool for analysing the workings of the complex geological world, making it possible to extract from this complexity a few simple underlying principles upon which an orderly understanding of geological processes can be based.

Many natural processes involve a flow of energy. The melting of an ice crystal, for example, requires energy to be transferred in the form of heat into the crystal from the 'surroundings' (the air or water surrounding the crystal). The crystal experiences an increase in its internal energy, which transforms it into liquid water. The process can be symbolized by writing down a formal reaction:

$$\underset{ice}{H_2O} \rightarrow \underset{water}{H_2O} \qquad (1.1)$$

in which molecules of water (H_2O) are represented as migrating from the solid state into the liquid state. Even at $0\,°C$, ice and water both possess internal energy associated with the individual motions of their constituent atoms and molecules. This energy content, which we can loosely visualize as heat 'stored' in each of the substances, is called the **enthalpy** (symbol H). Because the molecules in liquid water are more mobile than those in ice, the enthalpy of water (H_{water}) is greater than

1

that of an equivalent amount of ice (H_{ice}) at the same temperature. The difference can be written:

$$\Delta H = H_{water} - H_{ice} \qquad (1.2)$$

The Δ symbol (the Greek letter 'delta'), when written in front of H, signifies the difference in enthalpy between the initial (solid) and final (liquid) states of the compound H_2O. ΔH therefore symbolizes the amount of heat that must be supplied from the surroundings for the crystal to melt completely; this is called the **latent heat of fusion**, or more correctly the **enthalpy of fusion**, a quantity that can be measured experimentally or looked up in tables.

This simple example illustrates how one can go about documenting the energy changes that accompany geological reactions and processes, as a means of understanding why and when those reactions occur. This is the purpose of **thermodynamics**, a science that documents and explains quantitatively the energy changes in natural processes, just as economics analyses the exchange of money in international trade. Thermodynamics provides a fundamental theoretical framework for documenting and interpreting energy changes in processes of all kinds, not only in geology but in a host of other scientific disciplines ranging from chemical engineering to cosmology.

Thermodynamics, because it deals with very abstract concepts, has unfortunately acquired an aura of impenetrability in the eyes of many geology students, particularly those with less aptitude for mathematics. With this in mind, one objective of these opening chapters will be to show that thermodynamics, even at a simple and approachable level, can contribute a lot to our understanding of **chemical reactions and equilibrium** in the geological world.

Energy changes in chemical systems are most easily introduced by analogy with mechanical forms of energy that are familiar from school physics.

1.2 Energy in mechanical systems

The energy of a body is simply its capacity for doing work. 'Work' can take various forms, but usually implies the movement of a body from one position to another against some form of physical resistance (friction, gravity, electrostatic forces, etc.). Then:

$$\text{work done} = \begin{array}{c} \text{force required} \\ \text{to move body} \end{array} \times \begin{array}{c} \text{distance body} \\ \text{is moved} \end{array} \qquad (1.3)$$

$$\textit{joules} \qquad\qquad \textit{newtons} \qquad\qquad\quad \textit{metres}$$

So, for example, the work done in transporting a train-load of iron ore from A to B is the mechanical force required to keep the train rolling multiplied by the distance by rail from A to B. The energy required is obtained from the combustion of fuel in the engine.

One can distinguish two main kinds of mechanical energy. Firstly, an object can do work by means of its motion. A simple example is the use of a hammer to drive a nail into wood. Work is involved because the wood resists the penetration of the nail. The energy is provided by the downward-moving hammer head which, because of its motion, possesses **kinetic energy** (a name derived from the Greek *kinetikos*, meaning 'moving'). The kinetic energy of a body of mass m travelling with velocity v is given by

$$E_k = \tfrac{1}{2} \underset{\text{(kg)}}{m} \underset{\text{(m s}^{-1})^2}{v^2} \qquad (1.4)$$

$$\underset{\text{(J)}}{}$$

Thus the heavier the hammer (m) and/or the faster it moves (v), the more kinetic energy it possesses and the further in it will drive the nail. For similar reasons, a fast-moving stream can carry a greater load of sediment than a slow-moving one.

Secondly, an object in a gravitational field possesses energy by virtue of its position in that field and this is called **potential energy**. The water held behind a hydroelectric dam has a high potential energy: under the influence of the earth's gravitational field it would normally flow downhill until it reached sea level, but the dam prevents this happening. The fact that the controlled downward flow of this water under gravity can be made to drive turbines and generate electricity indicates that the water behind the dam has the potential for doing work, and therefore possesses energy.

The potential energy of an object of mass m at a height h above the ground is given by

$$\underset{\text{(J)}}{E_p} = \underset{\text{(kg)}}{m} \quad \underset{\text{(m s}^{-2})}{g} \quad \underset{\text{(m)}}{h} \qquad (1.5)$$

where g is the acceleration due to gravity ($9.81\,\text{m s}^{-2}$). (Similar equations can be written representing the potential energies of bodies in other types of force field, such as those in electric and nuclear fields.)

An important feature of potential energy is that its value as calculated from Equation 1.5 depends upon the baseline chosen for the measurement of height h. The potential energy calculated for an object in a second-floor laboratory, for example, will differ according to whether its height is measured from the laboratory floor, from the ground level outside, or from sea level. The last of these alternatives seems at first sight to be the most widely applicable standard to adopt, but even that reference point fails to provide a baseline which can be used for the measurement of height and potential energy down a deep mine (where both quantities would have negative values measured relative to sea level). This ambiguity forces us to recognize that potential energy is not something we can express on an absolute scale having a universal zero point, as we do in the case of temperature or electric charge. The value depends upon the 'frame of reference' we choose to adopt. We shall discover that this characteristic applies to chemical energy as well. It seldom presents practical difficulties because in thermodynamics one is concerned with energy changes, from which the baseline-dependent element cancels out (provided that the energy values used have been chosen in such a way that they all relate to the same frame of reference).

A body possesses kinetic and potential energy by virtue of its overall motion and position. There is also an internal contribution to its total energy from the individual motions of its constituent atoms and molecules, which are continually vibrating, rotating and – in liquids and gases – migrating about. This internal component, the aggregate of the kinetic and potential energies of all the atoms and molecules present, is what we mean by the **enthalpy** of the body. Enthalpy is closely related to the concept of heat (and was at one time referred to, rather misleadingly, as 'heat content'). Heat is one of the mechanisms by which enthalpy can be transferred from one body to another. The effect of heating a body is simply to increase the kinetic energy of the constituent atoms and molecules, and therefore to increase the enthalpy of the body as a whole.

Natural processes are continually converting energy from one form into another. One of the fundamental axioms of thermodynamics, known as the First Law, is that **energy can never be created, destroyed or 'lost'** in such processes, but merely changes its form (Box 1.1). Thus the energy given out by a reaction is matched exactly by the amount of energy gained by the surroundings.

1.3 Energy in chemical and mineral systems; free energy

Experience tells us that mechanical systems in the everyday world tend to evolve in the direction that leads to a **net reduction of total potential energy**. Water runs downhill, electrons arc drawn toward atomic nuclei, electric current flows from 'live' to 'neutral', and so on. The potential energy released by such changes reappears as other forms of energy or work: the kinetic energy of running water, the light energy radiated by electronic transitions in atoms (Chapter 6), the heat generated by an electric fire.

Thermodynamics visualizes chemical processes in a similar way. Reactions in chemical or geological systems arise from differences in what is called **free energy**, G, between products and reactants. The significance of free energy in chemical systems can be compared to that of potential energy in mechanical systems. A chemical reaction proceeds

---

Box 1.1

The First Law of Thermodynamics

The most fundamental principle of thermodynamics is that energy is never created, lost or destroyed. It can be transmitted from one body to another or one place to another, and it can change its identity between a number of forms (as for example when the potential energy of a falling body is transformed into kinetic energy, or when a turbine converts mechanical energy into electrical energy). But we never observe new energy being created from scratch, nor does it ever just disappear. Accurate energy book-keeping will always show that **in all known processes total energy is always conserved**. This cardinal principle is called the **First Law of Thermodynamics**. The energy given out by a reaction or process is matched exactly by the amount of energy gained by the surroundings.

Implicit in the First Law is the recognition that work is equivalent to energy, and must be accounted for in energy calculations. When a compressed gas at room temperature escapes from a cylinder, it undergoes a pronounced cooling, often to the extent that frost forms around the valve. (A similar effect occurs when you blow on your hand through pursed lips.) The cause of the cooling is that the gas has had to do work during escaping. It occupies more space outside the cylinder than when compressed inside it, and it must make room for itself by displacing the surrounding atmosphere. Displacing something against a resisting force (in this case atmospheric pressure) constitutes work, which the gas can only accomplish at the expense of its enthalpy. This is related directly to temperature, so that a drain on the gas's internal energy reserves becomes apparent as a fall in temperature.

A similar cooling effect may operate when certain gas-rich magmas reach high crustal levels or erupt at the surface. An example is kimberlite, a type of magma that commonly carries diamonds up from the mantle. Kimberlite penetrates to the surface from depths where the associated gases are greatly compressed, and the work that they do in expanding as the magma–gas system bursts through to the surface reduces its temperature; kimberlites found in sub-volcanic pipes (diatremes) appear to have been emplaced in a relatively cool state.

---

in the direction which leads to a **net reduction in free energy,** and the chemical energy so released reappears as energy in some other form – the electrical energy obtained from a battery, the light and heat emitted by burning wood, and so on.

What form does free energy take? How can it be calculated and used? These questions are best tackled through a simple example. Imagine a sealed container partly filled with water (Figure 1.1). The space not filled by the liquid takes up water vapour (a gas) until a certain pressure of vapour is achieved, called the **equilibrium vapour pressure** of water, which is dependent only upon the temperature (assumed to be constant). H_2O is now present in two stable forms, each one having physical properties and structure distinct from the other: these are called **phases**. From this moment on, unless circumstances change, the system will maintain a constant state, called **equilibrium**, in which the rate of evaporation from the liquid phase is matched exactly by the rate of condensation from the vapour phase: the relative volumes of the two phases will therefore remain constant.

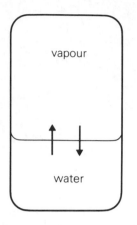

Figure 1.1 A simple model of chemical equilibrium between two coexisting phases, water and its vapour. The equilibrium can be symbolized by a simple equation.

$$H_2O \rightleftharpoons H_2O$$
liquid   vapour

At equilibrium, the migration of water molecules from the liquid to the vapour (evaporation, upward arrow) is balanced exactly by the condensation of molecules from vapour to liquid (downward arrow).

In this state of equilibrium the free energies associated with a given amount of water in each of these two phases must be equal. If that were not the case, a net flow of water molecules would begin from the high-G phase to the low-G phase, allowing the total free energy of the system to fall in keeping with the general tendency of chemical systems to minimize free energy. Any such flow, which would alter the relative proportions of the two phases, is inconsistent with the steady state observed. Clearly, at equilibrium, equivalent amounts of the two phases must have identical free energies:

$$G_{\text{vapour}} = G_{\text{liquid}} \tag{1.6}$$

This statement is in fact the thermodynamic definition of 'equilibrium' in such a system.

But here we seem to encounter a paradox. Common sense tells us that to turn liquid water in to vapour we have to supply energy, in the form of heat. The amount required is called the **latent heat of evaporation** (more correctly, the **enthalpy of evaporation**). This indicates that the vapour has a greater enthalpy (H_{vapour}) than an equivalent amount of the liquid (H_{liquid}):

$$H_{\text{vapour}} > H_{\text{liquid}} \tag{1.7}$$

The difference reflects the fact that water molecules in the vapour state have (a) greater potential energy, having escaped from the intermolecular forces that hold liquid water together, and (b) greater kinetic energy (owing to the much greater mobility of molecules in the gaseous state).

How can we reconcile Equations 1.6 and 1.7? Is it not common sense to expect the liquid state, in which the water molecules have much lower energies, to be intrinsically more stable than the vapour? What is it that sustains the vapour, in spite of its higher enthalpy, as a stable phase in equilibrium with the liquid?

The answer lies in the highly disordered state characteristic of the vapour. Molecules fly around in random directions, occasionally colliding but, unlike molecules in the liquid, free to disperse throughout the available volume. The vapour is said to possess a high **entropy** (S). Entropy is a parameter that records quantitatively the degree of internal disorder of a substance (Box 1.2). Entropy has immense significance in thermodynamics, owing to Nature's adherence to the Second Law of Thermodynamics. This states that all **spontaneous processes result in an increase of entropy**. The everyday consequences of the Second Law, which are so familiar that we often take them for granted, are discussed further in Box 1.2. In the present context, Nature's preference for disordered, high-entropy states of matter is what makes it possible for vapour to coexist with liquid. In a sense, the higher entropy of the vapour 'stabilizes' it in relation to the liquid state, compensating for the higher enthalpy required to sustain it.

Clearly, any analysis in energy terms, even of this simple example, will succeed only if the entropy difference (ΔS) between liquid and vapour is taken into account. This is why the definition of the free energy (or 'Gibbs energy') of each phase therefore incorporates an entropy term:

$$G_{\text{liquid}} = H_{\text{liquid}} - T.S_{\text{liquid}} \qquad (1.8)$$

$$G_{\text{vapour}} = H_{\text{vapour}} - T.S_{\text{vapour}} \qquad (1.9)$$

H_{liquid} and H_{vapour} are the enthalpies of the liquid and vapour respectively. S_{liquid} and S_{vapour} are the corresponding entropies. (Take care not to confuse the similar-sounding terms 'enthalpy' and 'entropy'.) The absolute temperature T, in kelvins, is assumed to be uniform in a system in equilibrium (Chapter 2), and therefore requires no subscript.

The important feature of these equations is the negative sign. It means that the vapour can have higher enthalpy (H) and higher entropy (S) than the liquid, and yet have the same free energy value (G), which must be true if the two phases are to be in equilibrium. Perhaps a more fundamental understanding of the minus sign can be gained by rearranging Equations 1.8 and 1.9 into this form:

$$H = G + T.S$$

The enthalpy of a phase can thus be seen as consisting of two contributions:

G The part that potentially can be released by the operation of a chemical reaction, which is logically called 'free' energy. This therefore provides a measure of the instability of a system (just as the potential energy of the water in a reservoir reflects its gravitational instability).

$T.S$ The part that is irretrievably bound up in the internal disorder of a phase at temperature T, and that is therefore not recoverable through chemical reactions.

Equations 1.8 and 1.9 express the fundamental contribution that disorder makes to the energetics of chemical and geological reactions, a question we shall take up again in the following sections.

UNITS

Enthalpy, entropy and free energy, like mass and volume, are classified as **extensive properties**. This means that their values depend on the amount of material present. (On the other hand temperature, density, viscosity and similar properties are said to be **intensive properties**, because they are unrelated to the size of the system.)

In published tables of enthalpy and entropy (Chapter 2) the values

7

Box 1.2

Some properties of entropy

The concept of disorder is of fundamental importance in thermodynamics, because it allows us to distinguish those processes and changes that occur naturally – 'spontaneous' processes – from those that do not. We are accustomed to seeing a cup shattering when it falls to the floor, but we never see the fragments reassemble themselves spontaneously to form a cup hanging on the dresser hook. Nor is it a natural experience for the air in a cold room to heat up a warm radiator. The direction of change that we accept as natural always leads to a more disordered state than we began with.

To apply such reasoning to the direction of chemical change, we need a variable that quantifies the degree of disorder in a chemical system. In thermodynamics we refer to this quantity as the **entropy** of the system. To define entropy rigorously lies beyond the scope of this book, but it is worth identifying the processes that lead to an increase of entropy. The entropy of a system depends on

(a) the distribution of matter or of individual chemical species in the system, and
(b) the distribution of energy.

Entropy and the distribution of matter

(a) Entropy increases as a substance passes from the solid state to the liquid state to the gaseous state:

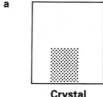

| **Crystal** | **Liquid/glass** | **Gas** |

Regular array of atoms/ molecules confined to their own part of system (ordered and touching).

Irregular (mobile) array of atoms/molecules confined to their own part of the system (disordered and touching).

Highly mobile atoms/ molecules scattered throughout the system (disordered and not touching).

(b) Entropy increases when a gas expands:

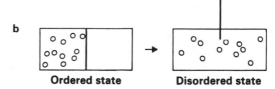

Ordered state   **Disordered state**

(c) Entropy increases when pure substances are mixed together:

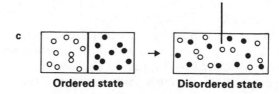

Ordered state   **Disordered state**

Entropy and the distribution of energy

The entropy of a system increases:

(d)   when a substance is heated

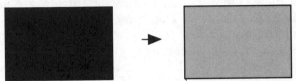

Key

hot

cold

because raising the temperature makes the random motions of atoms and molecules more vigorous.

(e)   when heat flows from a hot body (e.g. a radiator) to a cold body (a roomful of cold air):

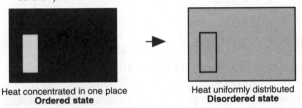

Heat concentrated in one place
Ordered state

Heat uniformly distributed
Disordered state

(f)   when chemical energy (fuel + oxidant) is transformed into heat:

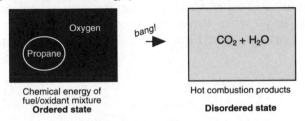

Chemical energy of
fuel/oxidant mixture
Ordered state

Hot combustion products

Disordered state

(g)   when mechanical energy is transformed into heat (e.g. by friction).

The Second Law of Thermodynamics

The Second Law states that the operation of any spontaneous process leads to an (overall) increase of entropy. Our experience of this law is so intricately woven into the fabric of everyday life that we are scarcely aware of its existence, but its impact on science is nonetheless profound. The expansion of a gas is a spontaneous process involving an increase of entropy: a gas never spontaneously contracts into a smaller volume. Water never runs uphill. Applying heat to an electric fire will never generate electricity. All of these impossible events, were they to occur, would bring about a reduction of entropy and therefore violate the Second Law.

Entropy is lowest when energy is concentrated in one part of a system. This is a characteristic all of the energy resources that we exploit: water retained behind a hydroelectric dam, chemical energy stored in a tank of gasoline or in a charged battery, nuclear energy in a uranium fuel rod, etc. Entropy is highest when energy is evenly distributed throughout the system being considered, and in such circumstances it cannot be put to good use. Spontaneous (entropy-increasing) changes are always accompanied by a degradation in the 'quality' of energy, in the sense that it becomes dispersed more widely and uniformly. For an entertaining discussion of these concepts the reader is referred to the classic book by Bent (1965), Chapters 1–4.

* A glass is a solid having the disordered structure of a liquid, but deprived of atomic mobility (no flow). Its entropy is intermediate between crystalline solid and liquid.

given are those for one mole, abbreviated in the SI system to 'mol', of the substance concerned (18 g in the case of water). One therefore speaks of **molar** enthalpy and entropy, and of molar free energy and molar volume as well. The units of molar enthalpy and molar free energy are joules per mole ($J\,mol^{-1}$); those of molar entropy are joules per kelvin per mole ($J\,K^{-1}\,mol^{-1}$). The most convenient units for expressing molar volume are $10^{-6}\,m^3\,mol^{-1}$ (which are the same as $cm^3\,mol^{-1}$, the units used in older literature).

In thermodynamic equations like 1.8, temperature is expressed in *kelvins* (K). Kelvins are equal in magnitude to °C but the scale begins at the absolute zero of temperature (-273.18 °C), not at the freezing point of water (0 °C). The SI units for pressure are pascals (Pa; see Appendix B).

1.3.1 Free-energy changes

For the reasons discussed above in relation to potential energy, the numerical values of G_{liquid} and G_{vapour} have no absolute significance. In considering whether water will evaporate or vapour will condense in specific circumstances, what concerns us is the change in free energy arising from the liquid-to-vapour 'reaction'. The first step in calculating free-energy changes is to write down the process concerned in the form of a chemical reaction, as in Equation 1.1. For the water/vapour equilibrium:

$$\underset{liquid}{H_2O} \rightleftharpoons \underset{vapour}{H_2O} \qquad (1.10)$$

The equilibrium symbol ($\rightleftharpoons$) represents a balance between two competing, opposed 'reactions' taking place at the same time:

$$\text{'Forward' reaction: } \underset{\substack{\text{left to right}}}{} \underset{reactant}{liquid} \rightarrow \underset{product}{vapour}$$

$$\text{'Reverse' reaction: } \underset{\substack{\text{right to left}}}{} \underset{product}{liquid} \leftarrow \underset{reactant}{vapour}$$

According to convention, the free-energy *change* for the forward reaction (ΔG) is written:

$$\Delta G = G_{products} - G_{reactants}$$
$$= G_{vapour} - G_{liquid}$$

Each G can be expressed in terms of molar enthalpy and entropy values obtained from published tables (Equations 1.8 and 1.9). Thus

$$\Delta G = (H_{vapour} - T.S_{vapour}) - (H_{liquid} - T.S_{liquid})$$
$$= (H_{vapour} - H_{liquid}) - T(S_{vapour} - S_{liquid})$$
$$= \Delta H - T.\Delta S \qquad (1.11)$$

In this equation ΔH is the heat input required to generate vapour from liquid (the latent heat of evaporation). In the context of a true chemical reaction, it would represent the **heat of reaction** (strictly the **enthalpy of reaction**). If ΔH for the forward reactions is negative, heat must be given out by the reaction, which is then said to be **exothermic** ('giving out heat'). A positive value implies that the reaction will proceed only if heat is drawn from the surroundings. Reactions that absorb heat in this way are said to be **endothermic** ('taking in heat'). ΔS represents the corresponding entropy change between liquid and vapour states.

The values of H_{vapour}, H_{liquid}, S_{vapour} and S_{liquid} can be looked up as molar quantities for the temperature of interest (e.g. room temperature $\simeq 298$ K) in published tables. In this case, ΔH and ΔS can be calculated by simple difference, leading to a value for ΔG (taking care to enter the value of T in kelvins, not Celsius). From the sign obtained for ΔG, it is possible to predict in which direction the reaction will proceed under the conditions considered. A negative value indicates that the products are more stable – have a lower free energy – than the reactants, so that the reaction can be expected to proceed in the forward direction. If ΔG is positive, on the other hand, the 'reactants' will be more stable than the 'products', and the reverse reaction will predominate. In either case, reaction will lead eventually to a condition where $\Delta G = 0$, signifying that equilibrium has been reached.

Now let us see how these principles apply to minerals and rocks.

1.4 Stable, unstable and metastable minerals

The terms 'stable' and 'unstable' have a more precise connotation in thermodynamics than in everyday usage. In order to grasp their meaning in the context of minerals and rocks, it will be helpful to begin by considering a simple physical analogue. Figure 1.2a shows a rectangular block of wood in a series of different positions relative to some reference surface, such as a table top. These configurations differ in their potential energy, represented by the vertical height of the centre of gravity of the block, shown as a dot, above the table top. Several general principles can be drawn from this physical system which will later help to illuminate some essentials of mineral equilibrium:

(a) Within this frame of reference, configuration D has the lowest potential energy possible, and one calls this the **stable** position. At the other extreme, configurations A and C are evidently **unstable**,

Figure 1.2 Potential energy of a rectangular wooden block in various positions on a planar surface. a. Four positions of the block, showing the height of its centre of gravity in each case. b. The pattern of potential-energy change as the block topples, for the unstable (A, C) and metastable (B) configurations.

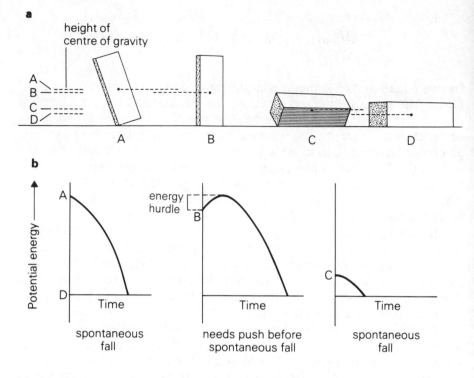

because in these positions the block will immediately fall over, ending up in position D. Both clearly have higher potential energy than D.

(b)   In discussing stable and unstable configurations, one need not consider all forms of energy possessed by the wooden block, some of which (for example, the total electronic energy) would be difficult to quantify. Mechanical stability depends solely upon the relative potential energies of, or energy differences between, the several configurations, and not on their absolute energy values.

(c)   Configuration B presents something of a paradox. It has a potential energy greater than the unstable state C, yet, if left undisturbed, it will persist indefinitely, maintaining the appearance of being stable. The introduction of a small amount of energy, such as a person bumping into the table, will however be sufficient to knock it over. The character of configuration B can be clarified by plotting potential energy against time as the block topples over (Figure 1.2b). For both unstable positions A and C, the potential energy falls continuously to the value of position D; but in the case of position B the potential energy must first rise slightly, before falling to the minimum value. The reason is that the block has to be raised on to its corner (position A) before it can fall over, and the work involved in so raising its centre of gravity constitutes a potential energy 'hurdle' which has to be surmounted before the block can topple. By inhibiting the spontaneous toppling of the block, this hurdle stabilizes configuration B. One uses the term **metastable** to describe any high-potential-energy state which is stabilized by such an energy hurdle.

The application of this reasoning to mineral stability can be illustrated by the minerals calcite and aragonite, whose ranges of stability in pressure–temperature space are shown in the form of a **phase diagram** in Figure 1.3a. These minerals are alternative crystallographic forms of calcium carbonate ($CaCO_3$), stable under different physical conditions. The phase diagram shown in Figure 1.3a is divided into two areas called **stability fields**, one representing the range of applied pressure and temperature under which calcite is the stable mineral; the other, at higher pressures, indicating the range of conditions favouring aragonite.

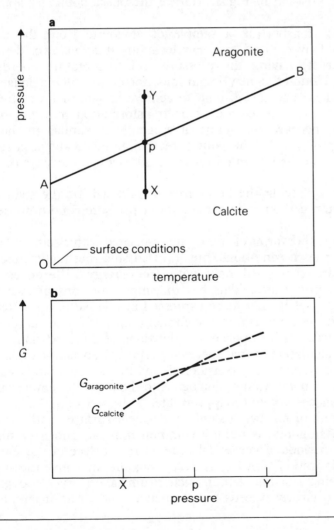

Figure 1.3 Stability of $CaCO_3$ polymorphs. a. Pressure–temperature phase diagram showing stability fields of calcite and aragonite. The line A–B is the phase boundary, indicating the P–T conditions under which calcite and aragonite are able to coexist in stable equilibrium. b. Free energy (G) of calcite and aragonite over a range of pressures along the isothermal line X–Y in the P–T diagram above.

The stability fields are separated by a line, called a **phase boundary**, which defines the restricted set of circumstances under which calcite and aragonite can coexist together **in equilibrium**.

The energetics of the calcite–aragonite system are illustrated in Figure 1.3b, which shows how the molar free energies of the two minerals vary along the line X–Y in Figure 1.3a. At high pressure (Y), deep within the crust, the molar free energy of aragonite is less than that of calcite, and thus aragonite is the **stable** mineral under these conditions, analogous to configuration D of the wooden block in Figure 1.2a. At a lower pressure (X) nearer to the surface, however, the position is reversed: calcite has the lower free energy and is therefore the stable mineral. The lines representing the free energy of calcite and aragonite as a function of pressure cross over in Figure 1.3b at a point marked p. Here the two minerals have equal molar free energies, and are therefore in chemical equilibrium with each other. Point p therefore marks the position in Figure 1.3b of the phase boundary appearing in Figure 1.3a.

Imagine transporting a sample of aragonite from the conditions represented by point Y to a new location (at a shallower depth in the Earth's crust) having the pressure and temperature coordinates of point X. Under the new conditions aragonite will no longer be the stable mineral and it will tend to achieve a state of lower free energy by recrystallizing to calcite. This transformation may not occur immediately, because the status of aragonite is similar to the wooden block in position B. The same three points made above in relation to Figure 1.2 can be reiterated for the calcite–aragonite system:

(a) Calcite, having the lower free energy, will be the stable form of calcium carbonate under the lower pressure conditions defined by X.

(b) Many other forms of energy are associated with calcite and aragonite under such conditions, but in discussing thermodynamic stability we are concerned only with **free-energy differences** between alternative states. This has the important consequence that free energy needs only to be expressed in relative terms, referred to a convenient but arbitrary common point, a sort of thermodynamic 'sea level'. All important applications of thermodynamics involve the calculation of free-energy differences between the various states of the system being considered, and the notion of an absolute scale of free-energy values, analogous to the absolute temperature scale, is unnecessary and inappropriate.

(c) In spite of not being stable under near-surface conditions (Figure 1.3), aragonite is quite a common mineral, and may survive for long periods of geological time on the surface of the Earth. Like configuration B in Figure 1.2a, aragonite may give the appearance of being a stable state in such cirumstances, even though its free energy clearly exceeds that of calcite. (Under certain circumstances, aragonite may actually crystallize under near-surface conditions: for example, the shells of planktonic gastropods (pteropods) are constructed of aragonite precipitated directly from sea water.) The explanation for this appearance of stability is that aragonite is stabilized by a free-energy 'hurdle', just like its mechanical analogue.

The free-energy path followed by aragonite (under the conditions shown by X in Figure 1.3a), as it undergoes transformation into calcite, is shown Figure 1.4. The energy hurdle exists because rearranging the crystal structure of aragonite into the form of calcite involves some work being done in breaking bonds and moving atoms about. Though this energy investment is recovered several times over in the net release of free energy as the reaction proceeds, its importance in determining whether the reaction can begin is considerable. In recognition of this influence, the height of the energy hurdle is called the **activation energy** of the reaction (symbol E_a).

Figure 1.4 illustrates an important distinction between two major domains of chemical science. **Thermodynamics** is concerned with the free-energy changes associated with chemical equilibrium between phases, and provides the tools for working out which mineral assemblages will be stable under which conditions. Only the initial and final states are of interest in thermodynamics, and attention is confined to *net* energy differences between reactants and products (ΔG, ΔH, ΔS), as Chapter 2 will show. The science of **chemical kinetics** deals with the mechanics of the reactions that lead to equilibrium, and the rates at which they occur. In this area, as we shall see in Chapter 3, the activation energy asserts a dominant rôle, accounting for the strong influence of temperature on many geological processes.

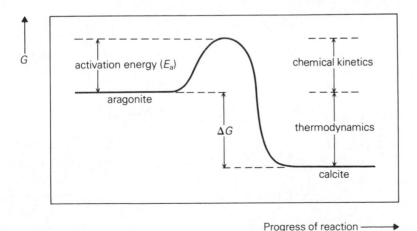

Figure 1.4 The free-energy path during the recrystallization of aragonite into calcite at the $P–T$ coordinates X in Figure 1.3.

1.5 Bibliography

Atkins, P. W. (1994) *Physical chemistry*, 5th edn, Oxford University Press, Oxford.

Bent, H. A. (1965) *The Second Law*, Oxford University Press, New York (Chapters 1–4 are particularly recommended.)

Smith, E. B. (1990) *Basic chemical thermodynamics*, 4th edn, Clarendon Press, Oxford.

2
Equilibrium in geological systems

2.1 The significance of mineral stability

Igneous and metamorphic rocks form in places which are not, on the whole, directly accessible to the investigating geologist. To discover how such rocks are produced within the earth, one must resort to indirect lines of enquiry. The most important clues consist of the minerals that the rocks themselves contain. A given mineral crystallizes as a stable phase only within restricted ranges of pressure and temperature, as we saw in the case of aragonite which occurs stably only at high pressures (Figure 1.3a). Subjecting the mineral to conditions that fall outside its stability range will cause another mineral that is stable under those conditions (like calcite at low pressure) to crystallize in its place. The stability of other types of minerals may depend in a similar way on the pressure of water vapour or some other gaseous component present during crystallization, and such a mineral will occur in a rock only if the vapour pressure present during its formation falls within the appropriate range.

The sensitivity of such minerals to the physical circumstances of their formation offers the petrologist tremendous opportunities because, when found in an igneous or metamorphic rock now exposed at the surface, they provide a means of establishing in quantitative terms the physical environment in which that rock originally crystallized. The study of mineral stability therefore offers the key to a veritable library of information, sitting in the rocks waiting to be utilized, about conditions and processes deep within the Earth's crust and upper mantle.

The usual way to establish the physical limits within which a mineral is stable – and beyond which it is unstable – is to cook it up in a laboratory experiment (Box 2.1). Technology today is capable of reproducing in the laboratory the physical conditions (temperature T, pressure P, water vapour pressure P_{H_2O}, and so on) encountered

Box 2.1

Phase equilibrium experiments with minerals

The process of mapping out a phase diagram like that in Figure 2.1 is illustrated in diagram a. Each of the filled and open circles represents an individual experiment in which a sample of the relevant composition is heated in a pressure-vessel at the pressure and temperature indicated by the coordinates, for a sufficient time for the phases to react and equilibrate with each other. At the end of each experiment – which may last hours, days or even months, depending on the time needed to reach equilibrium – the sample is 'quenched', meaning that it is cooled as quickly as possible to room temperature in order to preserve the phase assemblage formed under the conditions of the experiment (which on slower cooling might recrystallize to other phases – see Chapter 3). The sample is removed from its capsule, and the phase assemblage is identified under the microscope or by other methods. The symbol for each experiment is ornamented on the diagram in such a way that it indicates the nature of the phase assemblage observed, so that the results of a series of experiments allow the position of the phase boundary to be established. Conditions can be chosen for later experiments which allow accurate bracketing of its position in P–T space.

Experiments in the laboratory must necessarily be concluded in much shorter times than nature can afford to take to do the same job. Even at high temperatures, silicate reactions are notoriously sluggish, and the assemblage observed at the end of an experimental run may reflect an incomplete reaction or a metastable intermediate state rather than a true equilibrium assemblage. The proportions that this problem can sometimes assume are illustrated by the disagreement among the published determinations of the kyanite–sillimanite–andalusite triple point shown in diagram b. The present consensus places the triple point at about 4×10^8 Pa and $500\,°C$ (Figure 2.1).

One precaution that the experimenter can take is to ensure that the position of every phase boundary is established by approaching it from both sides, a procedure known as 'reversing the reaction'. In locating the kyanite–sillimanite phase boundary, for example, it is insufficient just to measure the temperature at which kyanite changes into sillimanite; the careful experimenter will also measure the temperature at which sillimanite, on cooling, inverts to kyanite.

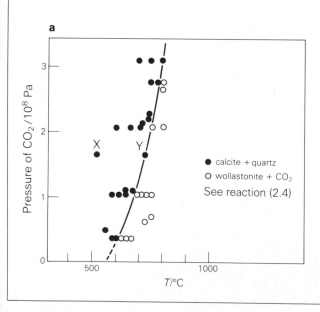

● calcite + quartz
○ wollastonite + CO_2
See reaction (2.4)

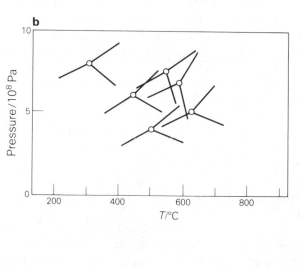

anywhere in the crust and in most of the upper mantle; practical details are given in the book by Holloway and Wood (1988). Experiments can be conducted in which minerals are synthesized at a series of accurately known temperatures and pressures, from starting materials of known composition. In each case, the minerals formed are considered to crystallize in **chemical equilibrium** with each other, or with a molten silicate liquid, and for this reason the experiments are called **phase-equilibrium experiments**. Experimental information like this is vital to the petrological interpretation of naturally occurring rocks. The results are used to map out, in phase diagrams similar to Figure 1.3a, the areas within which particular mineral assemblages are stable. In crossing the boundary from one stability field into a neighbouring one as physical conditions change, one mineral assemblage recrystallizes into another, a chemical reaction between coexisting minerals which transforms an unstable – or metastable – assemblage into a new, stable one. These boundaries, like the diagonal line in Figure 1.3a, are called **phase boundaries** or, more generally, **reaction boundaries**.

Phase diagrams are part of the language of petrology, a powerful means of portraying and interpreting the phase-equilibrium data relevant to igneous and metamorphic petrogenesis. Reading and interpreting such diagrams, in the light of underlying thermodynamic principles, is one of the basic skills essential to every geologist.

Just as atmospheric pressure represents the weight of the column of air above us under the influence the Earth's gravitational field, so the pressure experienced by a rock buried at some depth within the Earth reflects the total weight of the column of rock + ocean + atmosphere resting upon it. The lithostatic pressure P therefore increases with depth d in a simple manner (see Figure 2.2) conveniently approximated as:

$$P \approx 3 \times d$$

where P is expressed in units of $10^8\,Pa$ and d is in km. The high pressure applied in some phase-equilibrium experiments is simply the means by which we imitate the effect of depth.

2.2 Systems, phases and components

To avoid confusion, one must be clear about the meaning of several terms which are used in a specialized sense in the context of phase equilibrium.

SYSTEM
This is a handy word for describing any part of the world to which we wish to confine attention. Depending on the context, **system** could mean the whole of the Earth's crust, or the oceans, or a cooling magma

chamber, or an individual rock, or a sample undergoing a phase-equilibrium experiment. In most cases the term will refer to a collection of geological phases (see below) interacting with each other.

An **open system** is one that is free to exchange both matter and energy with its surroundings. The sea is an open system; it may be useful to consider it separately for the purpose of discussion, but one must recognize that it receives both sunlight and river water from outside, and loses heat, water vapour and sediment to the atmosphere and crust.

A **closed system** is one that is sealed with respect to the transfer of matter, but that can still exchange energy with the surroundings. A sealed magma chamber would be a good example, its only interaction with its environment being the gradual loss of heat. An **isolated system** is one capable of exchanging neither mass nor energy with its surroundings, a notion of little relevance to the real world.

'System' may alternatively be used to denote a domain of chemical (rather than physical) space. Petrologists use the word to distinguish a particular region of **'compositional space'** to which attention is to be confined. Thus one speaks of 'the system $MgO-SiO_2$', referring to the series of compositions that can be generated by mixing these two chemical components in all possible proportions. The system so designated includes various minerals whose compositions lie within this range (the silica minerals, and olivine and pyroxene).

PHASE

The meaning of **phase** in physical chemistry and petrology is easy to grasp but cumbersome to put into words. In formal terms, a phase can be defined as 'a part or parts of a system occupying a specific volume and having **uniform physical and chemical characteritics** which distinguish it from all other parts of the system'.

Each individual mineral in a rock thus constitutes a separate phase, but that is not the end of the story. A sample of basalt phase, but that is not the end of the story. A sample of basalt collected from a solidified lava flow might be found to consist of four minerals, say plagioclase, augite, olivine and magnetite. But the igneous texture tells us that these minerals crystallized from a melt, which itself had uniform physical and chemical properties distinct from each of the crystalline minerals present. So in considering the phase relations that dictated the present character of the rock we must count the melt as a phase too, though it is now no longer present as a constituent. If the basalt is **vesicular** (that is, it contains empty bubbles or 'vesicles'), we have evidence that a sixth phase, water vapour, was also present as the rock formed.

(Note that water can be present as a dissolved species within a melt or a hydrous crystalline mineral. Being accommodated within the volume occupied by that phase, it does not then count as a phase in its own right. Only when separate bubbles of water vapour appear can one accord water the status of a separate phase. When this happens, as with the vesiculating lava, it is a sign that all other phases present contain as much water as they can accommodate, and the system is **saturated** with water.)

One must therefore be careful not to overlook additional phases which, though no longer present in a particular rock, may have influenced its formation or development. Here are some examples where this could happen:

(a) Some igneous rocks show textural evidence for the existence of two distinct (presumably immiscible) silicate liquids that once existed together in mutual equilibrium.

(b) A metamorphic rock may have developed in the presence of a vapour phase, permeating the grain boundaries between crystals, of which no visible trace now remains.

(c) Mineral veins are deposited from a fluid phase preserved only in occasional microscopic inclusions within crystals (Box 4.5).

(d) When a magma is produced by melting deep inside the Earth, it may be in equilibrium with minerals quite different from those that crystallize from the same melt at the surface.

Thus the minerals we see in a rock today may represent only a part, or even no part at all, of the original phase equilibria to which the rock owes its present constitution.

It is usual to refer to solid phases by the appropriate mineral name – quartz, kyanite, olivine, etc. The essential distinction between them is crystallographic structure, not their chemical composition (which several minerals may have in common).

COMPONENT

The basic chemical constituents of a system, of which the various phases are composed, are called its **components**. The concept of a component is defined in a rather roundabout way: 'the components of a system comprise the minimum number of chemical (atomic and molecular) species required to specify completely the compositions of all the phases present'.

Consider a crystal of olivine, which at its simplest consists of the elements magnesium (chemical symbol Mg), iron (Fe), silicon (Si) and oxygen (O). One way to define the components of the olivine would be to regard each chemical element as a separate component because the composition of any olivine can be stated in terms of the concentrations of four elements:

$$Mg \quad Fe \quad Si \quad O$$

However, defining the components in this way fails to recognize an important property of all silicate minerals, including olivine: that the oxygen content is not an independent quantity, but is tied by valency

21

(Chapter 6, p. 134) to the amounts of Mg, Fe and Si present, being just sufficient to generate the oxides of each of these elements (this is explained in Box 8.3, p. 183). So, in describing the composition of an olivine the same information can be conveyed more economically in terms of the concentrations of only three components:

$$MgO \qquad FeO \qquad SiO_2$$

By using a property specific to olivines, however, a still more economical statement of olivine composition can be devised. The crystal chemistry of olivine (see Chapter 8) requires an olivine composition to conform to a general formula which we can represent by X_2SiO_4. X represents a type of atomic site in the olivine crystal structure which accommodates either Mg or Fe, but not Si. For every Si atom in the olivine structure, there have to be two divalent atoms present, each of which can be either Mg or Fe. Another way to symbolize this constraint is to write the formula as $(Mg,Fe)_2SiO_4$, in which '(Mg,Fe)' represents an atom of either Mg or Fe. One can now express the composition of an olivine as a combination of just two components:

$$Mg_2SiO_4 \qquad Fe_2SiO_4$$

The mineralogist calls these components the 'end-members' of the olivine series, and gives them the names forsterite (Fo) and fayalite (Fa) respectively.

In analysing the arithmetic of chemical equilibrium between minerals it is important to formulate the components of a system in such a way as to minimize their number, as the definition implies. What constitutes the minimum number depends on the nature of the system. In an experiment involving the melting of an olivine on its own, the composition of the melt, though different from the solid, still conforms to the olivine formula X_2SiO_4. The compositions of both phases present, olivine and melt, can therefore be expressed as proportions of only two components, Mg_2SiO_4 and Fe_2SiO_4 (Box 2.4). Systems consisting of only two components are called **binary** systems.

If, however, olivine coexists with, let us say, orthopyroxene, the formulation of components becomes less straightforward. Orthopyroxenes are composed of the same four elements as olivine, but they combine in different proportions. The general formula of orthopyroxene, $X_2Si_2O_6$, reveals an X : Si ratio (1 : 1) lower than for olivine (2 : 1). The composition of a pyroxene cannot therefore be expressed in terms just of the two olivine end-members. To represent the separate compositions of olivine and orthopyroxene in this system three components will be needed:

Either: $\quad Mg_2SiO_4 \quad Fe_2SiO_4 \quad SiO_2$

Or: $\qquad\quad MgO \qquad\;\; FeO \qquad SiO_2$

The identity of the components is less important here than their number. A system like this requiring three components to express all possible compositions is said to be **ternary**.

There are circumstances in which four components would be necessary, such as when olivine coexists with metallic iron (for example in certain meteorites). The amount of oxygen present is no longer determined solely by the metals present, as it would be in a system consisting entirely of silicates. One cannot express the composition of metallic iron as a mixture of oxides, so one must resort to using four components, Mg, Fe, Si and O, in order to describe all possible compositions in this **quaternary** system.

In this book, the general practice will be to refer to components by means of their chemical formulae. This will avoid confusion between phases and components, which can arise when a phase (for example the mineral quartz) happens to have the same chemical composition as one of the components (SiO_2) in the same system. However, in other literature it is quite common for end-member names to be used in this way as well (for example, 'forsterite' for Mg_2SiO_4).

2.3   Equilibrium

It is useful to distinguish between two aspects of equilibrium, thermal equilibrium and chemical equilibrium.

THERMAL EQUILIBRIUM
This simply means that all parts of the system have the same temperature: heat flowing from one part, A, to another part, B, is exactly balanced by the heat passing from part B to part A. There is no net transfer of heat.

CHEMICAL EQUILIBRIUM
This means that the distribution of components among the phases of a system has become constant, showing no net change with time. This steady state does not mean that the flow of components from one phase to another has ceased: equilibrium is a dynamic process. An olivine suspended in a magma is constantly exchanging components with the melt. At melt temperatures, atoms will diffuse across the crystal boundary, both into the crystal and out of it into the liquid. If the diffusion rates of element X in and out of the crystal are unequal, there will be a net change of the composition of each phase with time, a condition known as **disequilibrium**. Such changes usually lead eventually to a situation where, for every element present, the flux of atoms across the crystal boundary is the same in both directions, resulting in zero net flow, and no change of composition with time. This is what we mean by **equilibrium**.

The rate at which equilibrium is achieved varies widely and, as

Chapter 3 will show, disequilibrium is found to be a common condition in geological systems, particularly at low temperatures.

2.3.1 The Gibbs Phase Rule

A natural question to ask is how many phases can be in equilibrium with each other at any one time. In Figure 1.3a we looked at a simple system in which only two phases occurred. Most actual rocks, however, are nothing like as simple. What determines the mineralogical complexity of a natural rock? Which aspect of a chemical equilibrium controls the number of phases that participate in it?

This question was addressed in the 1870s by J. Willard Gibbs, the pioneer of modern thermodynamics. The outcome of his work was a simple but profoundly important formula called the Phase Rule, which expresses the number of phases that can coexist in mutual equilibrium (ϕ) in terms of the number of components (C) in the system and another property of the equilibrium called the **variance** (F). The Phase Rule can be stated symbolically as:

$$\phi + F = C + 2 \qquad (2.1)$$

The variance is sometimes known as the **number of degrees of freedom** (hence the symbol F). The concept is most easily introduced through an example. Figure 2.1 illustrates the phase relations between the minerals kyanite, sillimanite and andalusite. These minerals are all aluminium silicate polymorphs of identical composition. A single component (Al_2SiO_5) is therefore sufficient to cover the composition 'range' of the entire system.

Points A, B and C are three different points in 'P–T space'; they

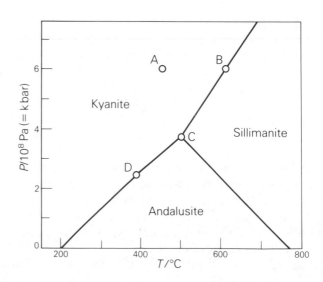

Figure 2.1 A P–T diagram showing phase relations between the aluminium silicate minerals (composition Al_2SiO_5). The pressure axis is graduated in units of 10^8 pascals, equal in magnitude to the traditional pressure units, kilobars (1 kbar = 10^3 bars = 10^8 Pa).

Kyanite is a triclinic mineral that usually occurs as pale blue blades in hand specimens. Sillimanite is orthorhombic and is commonly fibrous or prismatic in habit. Andalusite is also orthorhombic and is characteristically pink in hand specimens.

represent three classes of equilibrium which can develop in the system. The obvious difference between them is the nature of the equilibrium assemblage. Point A lies inside a field where only one phase, kyanite, is stable. Point B lies on the phase boundary between two stability fields, where two minerals, kyanite and sillimanite, are stable together. Point C, at the **triple point** where the three stability fields (and the three phase boundaries) meet, represents the only combination of pressure and temperature at which all three phases can exist stably together.

It is clear that the three-phase assemblage (kyanite + sillimanite + andalusite), when it occurs, indicates very precisely the state of the system (that is, the values of P and T) in which it is produced, because there is only one set of conditions under which this assemblage will crystallize in equilibrium. Using the Phase Rule (Equation 2.1), one can calculate that the variance F at point C is zero:

Point C	$\phi = 3$	(3 phases, ky + sill + andal)
	$C = 1$	(1 component, Al_2SiO_5)
	$3 + F = 1 + 2$	
Therefore	$F = 0$	an *invariant* equilibrium

A variance of zero means that the three-phase equilibrium assemblage completely constrains the state of the system to a particular combination of P and T. Such a situation is called **invariant**. There is no latitude (no degree of freedom) for P or T to change at all, if the assemblage is to remain stable. Any such variation would lead to the disappearance of one or more of the three phases, thus altering the character of the equilibrium. So the discovery of this mineral assemblage in a metamorphic rock (a rare event, since only about a dozen natural occurrences are known) ties down very precisely the conditions under which the rock must have been formed, assuming that:

(a)   the kyanite–andalusite–sillimanite assemblage represents an **equilibrium state** obtaining as the rock formed, and not simply an uncompleted reaction from one assemblage to another, and

(b)   the $P–T$ coordinates of the invariant point in Figure 2.1 are accurately known from experimental studies. (Whether this requirement is satisfied in the case of the Al_2SiO_5 polymorphs is arguable – see Box 2.1 – but in the present discussion we shall ignore this difficulty.)

The two-phase equilibrium between, say, kyanite and sillimanite is less informative. The coexistence of these minerals indicates that the state of the system in which they crystallized must lie somewhere on the kyanite–sillimanite phase boundary, but exactly where along this line remains uncertain unless we can specify one of the coordinates of point B. We only need to specify one coordinate (for example, temperature) because the other is then fixed by intersection of the specified coordinate with the phase boundary. According to the Phase Rule, the variance at B is equal to 1:

Point B	$\phi = 2$	(2 phases, ky + sill)
	$C = 1$	(1 component, Al_2SiO_5)
	$2 + F = 1 + 2$	
Therefore	$F = 1$	a *univariant* equilibrium

The one degree of freedom indicates that the state of the system is only unconstrained in one direction in Figure 2.1, along the phase boundary. One more piece of information is required (either T or P) to tie down the state of the system completely. The coexistence of kyanite and sillimanite in a rock will pinpoint the exact conditions of origin only in conjunction with other information about P or T.

At point A, where kyanite occurs alone, the variance is equal to 2:

Point A	$\phi = 1$	(1 phase, ky)
	$C = 1$	(1 component, Al_2SiO_5)
	$1 + F = 1 + 2$	
Therefore	$F = 2$	a *divariant* condition

Within the bounds of the **divariant** kyanite field, therefore, P and T can vary independently (two degrees of freedom) without upsetting the equilibrium phase assemblage (just kyanite). The one-phase assemblage is therefore little help in establishing the precise state of the system, because it leaves two variables (P and T) still to be specified.

The variance cannot be greater than 2 in a one-component system like Figure 2.1. In the more complex systems we shall meet in the following section, the phases present may consist of different proportions of several components. A complete definition of the state of such a system must then include the compositions of one or more phases, in addition to values of P and T. Such compositional terms (X_a X_b, etc., representing the mole fractions of a, b, etc. in a phase) contribute to the total variance, which can therefore, in multi-component systems, adopt values greater than 2.

Variance can be summarized in the following way. The 'state' of a system – whether we consider a simple experimental system or a real metamorphic rock in the making – is defined by the values of certain key intensive variables, including pressure (P), temperature (T) and, in multi-component systems, the compositions (Xs) of one or more phases. For a given equilibrium between specific phases, some of these values are automatically constrained in the phase diagram by the equilibrium phase assemblage. The variance of this equilibrium is the number of the variables that remain free to adopt arbitrary values, and that must be determined by some other means if the state of the system is to be defined completely.

2.4 Phase diagrams in $P-T$ space

The need to represent phase equilibrium data in visual form on a two-dimensional page leads to the use of various forms of phase diagram, each having its own merits and limitations. We begin by looking at $P-T$ diagrams.

The two phase diagrams so far considered (Figures 1.3a and 2.1) both show the effects of varying pressure and temperature on a system consisting of only one component ($CaCO_3$ or Al_2SiO_5). Other important examples of such **unary** systems are discussed in Box 2.2.

$P-T$ diagrams can also be used to show the pressure–temperature characteristics of multi-component reactions and equilibria. An example is shown in Figure 2.2. The univariant boundary in this diagram represents not a phase transition between different forms of the same compound, but a reaction or equilibrium between a number of different compounds:

$$NaAlSi_2O_6 + SiO_2 \rightleftharpoons NaAlSi_3O_8 \qquad (2.2)$$

jadeite   *quartz*   *albite*

(a pyroxene)   *(a feldspar)*

For this reason the term **reaction boundary** or (**equilibrium boundary**) is used. It marks the $P-T$ threshold across which reaction occurs, or the conditions at which univariant equilibrium can be established.

Two components are sufficient to represent all possible phases in this system. We can choose them in a number of equivalent ways; selecting $NaAlSi_2O_6$ and SiO_2 is as good a choice as any. Applying the Phase Rule to point X:

Point X	$\phi = 2$	(2 phases, jadeite + quartz)
	$C = 2$	(2 components)
	$3 + F = 2 + 2$	
Therefore	$F = 2$	a *divariant* field

27

Box 2.2

Other one-component phase diagrams

Graphite-Diamond

The phase relations between the two crystalline forms of carbon (Chapter 8) are shown in diagram A. Note the very high pressure (in the region of $20 \times 10^8\,Pa$) required to stabilize diamond. For this reason, it can only form deep inside the Earth's mantle (an equivalent depth scale is shown at the right-hand side). Moreover, the minimum pressure increases with temperature, so still higher pressures are necessary to stabilize diamond in the hot interior of the Earth than would be the case at room temperature. The curve marked 'geotherm' shows how temperature increases with depth beneath ancient continental shields (where diamond-bearing kimberlites are found). From the point at which the geotherm enters the diamond stability field, it is clear that diamonds can only be formed at depths greater than about $120\,km$ ($\approx 40 \times 10^8\,Pa$).

Each stability field in diagram A is divariant, and the boundary between them is univariant. Because there are only two polymorphs of carbon (i.e. $\phi < 3$), there is no invariant point (i.e. $F > 0$) in this phase diagram.

The H_2O phase diagram

Diagram B shows the phase equilibria between the familiar forms of pure H_2O as a function of pressure and temperature. Note that the axes are not drawn to scale. Vapour, liquid and solid coexist at only one point in the diagram ($0.06 \times 10^5\,Pa$ and $0.008\,°C$). This **triple point** lies well below atmospheric pressure $P_A = 1 \times 10^5\,Pa$ (dashed line). The curve T–C shows the vapour pressure at which liquid water and vapour are in mutual equilibrium (the **equilibrium** or **saturation vapour pressure**) as a function of temperature. Under such conditions, the vapour is said to be saturated; at pressures below the phase boundary, however, the vapour is unsaturated and no liquid water can form. The equilibrium vapour pressure curve T–C rises with temperature to reach P_A at $100\,°C$. We can define the **boiling point** of pure water (T_b) as the temperature at which equilibrium vapour pressure (the univariant curve T–C) becomes equal to atmospheric pressure. The vapour then exerts sufficient pressure to displace its atmospheric surroundings and form bubbles in the liquid, the everyday phenomenon of boiling. (Note that if the atmosphere pressure is lowered below P_A, as on Mount Everest, for example, water will boil at temperatures lower than $100\,°C$.)

At room temperature ($25\,°C$), the equilibrium vapour pressure of water lies well below atmospheric pressure P_A. Atmospheric water vapour can be considered to exert a **partial vapour pressure** in proportion to its concentration in the air. If this partial pressure is below the saturation vapour pressure, there will be a net evaporation of water to vapour (clothes dry, puddles evaporate), whereas if the partial pressure of water exceeds the equilibrium vapour pressure, water vapour will tend to condense to liquid water (dew, mist, rain). The **relative humidity** expresses the actual water vapour pressure in a given body of air as a percentage of the saturation vapour pressure at the temperature concerned.

The univariant curve dividing the liquid and vapour fields ends abruptly at invariant point C, known as the **critical point** of water. At this combination of P and T, the structural distinction between liquid and gaseous states vanishes. The two states merge into a single phase. At higher temperatures and pressures, H_2O exists as a homogeneous single phase called a **supercritical fluid**, in which are combined the properties of a highly compressed gas and a superheated liquid. Some of the hydrothermal fluids responsible for depositing ore bodies come into this category, and the noun 'fluid' used alone often has this connotation in geology. All liquid/gas systems become supercritical fluids under sufficiently extreme conditions.

The atmospheric pressure (P_A) 'isobar' cuts the ice–water phase boundary at exactly $0\,°C$ (T_m). This phase boundary has the unusual property of a *negative slope*, a feature upon which every ice-skater unconsciously depends. It expresses the fact that the melting point of ice decreases as pressure is increased, so that ice close to $0\,°C$ can be melted simply by the application of pressure, such as the skater's weight acting on the narrow runner of the skate. This behaviour, like many other properties of water (Box 4.1), is unique among the common liquids: melting points for most other materials rise with increasing pressure, as the phase diagram for carbon dioxide (inset in diagram B) illustrates. See Exercise 2 at end of this chapter.

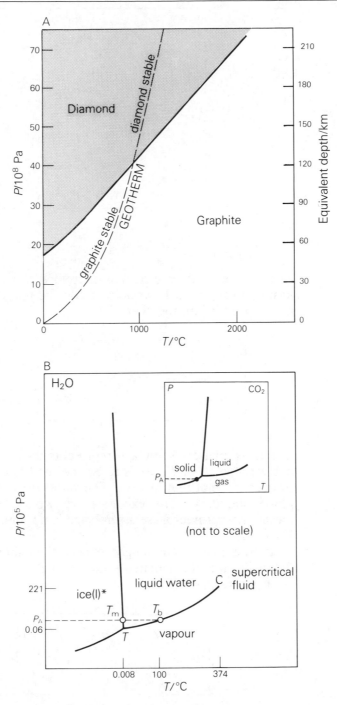

*Ice (I) is one of 8 structural forms of ice stable at different pressures and temperatures.

Figure 2.2 *P–T* diagram showing the experimentally determined reaction boundary (solid line) for the reaction

jadeite + quartz → albite

Jadeite and albite are both aluminosilicates of sodium (Na). The depth scale on the right-hand margin indicates the correspondence between pressure and depth within the Earth.

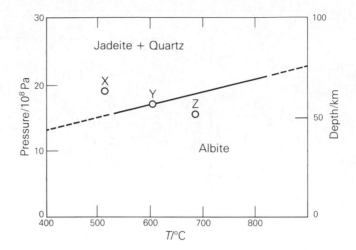

The jadeite + quartz field is therefore a divariant field like that of kyanite in Figure 2.1. At point Y, on the phase boundary, three phases are in equilibrium together:

Point Y $\phi = 3$ (3 phases, jadeite + quartz + albite)
                  $C = 2$ (2 components)
            $3 + F = 2 + 2$
Therefore $F = 1$ a *univariant* equilibrium

The three-phase assemblage represents a univariant equilibrium: only one variable, P or T, needs to be specified to determine completely the physical state of they system. The value of the other can be read off the phase boundary. The existence of two components in this system means that a three-phase assemblage is no longer invariant, as it was in Figure 2.1.

At first glance one might expect the albite field (point Z) to be divariant like the jadeite + quartz field, but here the Phase Rule springs a surprise:

Point Z $\phi = 1$ (1 phases, albite)
                  $C = 2$ (2 components)
            $1 + F = 2 + 2$
Therefore $F = 3$ a *trivariant* condition

Analysing the albite field in this way, it appears necessary to specify the values of three variables to define the state of the system in this condition. P and T account for two of them, but what can be third variable be? The answer becomes clear if we ask what requirements must be met if, in passing from X to Z, we are to generate albite alone. If the mixture of jadeite and quartz contains more molecules of SiO_2 than $NaAlSi_2O_6$, a certain amount of quartz will be left over after all the jadeite has been used up. The resultant assemblage at Z will therefore be albite + quartz. The presence of two phases leads to a variance of 2 for this field, as originally expected. Conversely, if we react SiO_2 with an excess of $NaAlSi_2O_6$ molecules, the resultant assemblage at Z would consist of albite + jadeite, again a divariant assemblage. The only way to form the one-phase assemblage, albite alone, is to combine jadeite and quartz in exactly equal molecular proportions, so that no quartz or jadeite is left over. In other words, to generate just albite in passing from X to Z we must control not only P and T but also a **compositional** property – the $NaAlSi_2O_6 : SiO_2$ ratio – of the system. This compositional requirement is the unsuspected third degree of freedom whose existence the Phase Rule has uncovered.

$P–T$ diagrams provide useful vehicles for portraying metamorphic conditions in the crust during mountain building (Yardley, 1989) and melting processes in the mantle (Box 2.5).

$P_V–T$ DIAGRAMS

Reactions in which all the reactants and products are crystalline minerals such as those illustrated in Figures 1.3a, 2.1 and 2.2 are known as 'solid–solid reactions'. No vapour is involved, and its presence or absence in the experiments is immaterial to the equilibrium finally obtained (although it can accelerate progress toward equilibrium). Figure 2.3 illustrates another important class of reactions, in which a volatile constituent plays an essential rôle.

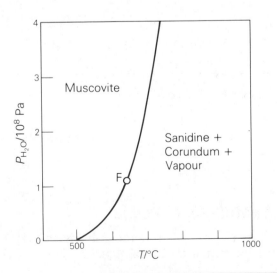

Figure 2.3 $P_{H_2O}–T$ diagram showing the 'dehydration curve' of the mica muscovite, showing the $P_{H_2O}–T$ conditions at which it breaks down into the assemblage sanidine (an alkali feldspar) + corundum + vapour. Muscovite and sanidine are aluminosilicates of potassium (K).

Figure 2.3 shows a reaction involving water (the dehydration of muscovite at high temperatures):

$$KAl_3Si_3O_{10}(OH)_2 \rightleftharpoons KAlSi_3O_8 + Al_2O_3 + H_2O \qquad (2.3)$$

muscovite   sanidine   corundum   vapour
(a mica)   (a feldspar)

and diagram A in Box 2.1 shows a similar reaction involving carbon dioxide, which is important in the metamorphism of siliceous limestones:

$$CaCO_3 + SiO_2 \rightleftharpoons CaSiO_3 + CO_2 \qquad (2.4)$$

calcite   quartz   wollastonite   vapour
(a pyroxene-like
mineral)

Because molecules of H_2O and CO_2 are involved in these reactions, vapour pressure exerts a strong influence on the position of equilibrium.

The experiments from which these diagrams were prepared were carried out in the presence of an excess of H_2O or CO_2 respectively (as a separate gas phase), so that at all times the experimental charges were saturated with the volatile constituent. The consequent existence of a vapour phase in all the experiments means that the vapour pressure was equal to the total pressure applied to the specimen. The general symbol for **vapour pressure** is P_V; one can use the more specific symbols P_{H_2O} or P_{CO_2} for the vapour pressure of water and carbon dioxide respectively.

Figure 2.3 is therefore a vapour pressure–temperature diagram, which can be analysed using the Phase Rule in the same way as Figure 2.2. For example, if we choose the components carefully in Figure 2.3 (Equation 2.3), we find that only three are necessary to constitute all four phases:

$$KAlSi_3O_8 \qquad Al_2O_3 \qquad H_2O$$

Point F	$\phi = 4$	(4 phases)	
	$C = 3$	(3 components)	
	$4 + F = 3 + 2$		
Therefore	$F = 1$	a *univariant* equilibrium	

2.4.1   Le Chatelier's principle

Behind the empirical facts of mineral stability, as represented by the experimentally determined $P-T$ diagrams in Figures 2.1, 2.2 and 2.3, there are some important thermodynamic principles which will help in the interpretation of phase diagrams.

The first of these concerns the distribution of phases in a phase diagram. Why is kyanite stable at high pressures, whereas andalusite can survive only at low pressure (Figure 2.1)? What properties of the two minerals dictate this behaviour? How is it that sillimanite is more stable than either of them in the highest temperature range?

The answers to these questions lie in a simple principle enunciated by H. L. Le Chatelier in 1884: when a system at equilibrium experiences a change in physical conditions, the system will adapt in a direction which tends to nullify the change. In the present context, the 'physical conditions' referred to are pressure and temperature.

Consider a system comprising kyanite and andalusite in mutual equilibrium, for example under the conditions represented by point D in Figure 2.1:

$$\underset{andalusite}{Al_2SiO_5} \rightleftharpoons \underset{kyanite}{Al_2SiO_5} \qquad (2.5)$$

How will the equilibrium respond if we attempt to increase the applied pressure (without changing the temperature)? Le Chatelier's principle suggests that the assemblage will adjust into a more compact form, because by taking up less space it can relieve the additional pressure applied. The system can accomplish this by recrystallizing andalusite (density $3.2 \, kg \, dm^{-3}$) into the denser polymorph kyanite (density $3.6 \, kg \, dm^{-3}$). By allowing the proportion of kyanite to increase at the expense of andalusite, the system can for the time being prevent any increase in pressure. Eventually, however, the andalusite is exhausted and the pressure, no longer constrained by univariant equilibrium, is able to rise into the kyanite field. From Le Chatelier's principle one can therefore show that in any $P-T$ phase diagram the higher-density (lower molar volume) phase assemblage will be found on the higher-pressure side of a reaction boundary. Diamond and liquid water are other examples (Box 2.2).

A second consequence of Le Chatelier's rule is that the phase assemblage on the high-temperature side of an equilibrium boundary is invariably the one having the higher enthalpy (Box 4.2).

Le Chatelier's principle

2.4.2 The Clapeyron equation

A second useful application of thermodynamic data to phase diagrams is to estimate the gradient (slope) of an equilibrium boundary in $P-T$ space. In the symbolism of calculus (see 'Differentiation' on p. 278, Appendix B), this is written $\dfrac{dP}{dT}$, meaning the rate at which P increases for a given increase of T as one follows the univariant boundary. The gradient has a sign (positive or negative) and a numerical magnitude (indicating whether it is gentle or steep).

Consider the reaction of Equation 2.2 and Figure 2.2:

$$\text{jadeite} + \text{quartz} \rightleftharpoons \text{albite}$$

The free-energy change of this reaction is:

$$\begin{aligned} \Delta G &= G_{\text{products}} - G_{\text{reactants}} \\ &= G_{\text{albite}} - (G_{\text{jadeite}} + G_{\text{quartz}}) \end{aligned} \tag{2.6}$$

The molar free energy of each phase (which could be calculated from tables of molar enthalpy and entropy) varies with pressure and temperature. ΔG therefore varies systematically across $P-T$ diagrams. The equilibrium boundary in Figure 2.2 marks the locus of $P-T$ coordinates for which $\Delta G = 0$. It can be shown (using quite simple calculus) that the condition for remaining on the univariant equilibrium boundary as P and T are varied by small amounts dP and dT is that:

$$\frac{dP}{dT} = \frac{\Delta S}{\Delta V} \tag{2.7}$$

where ΔS and ΔV are the entropy and volume changes occurring during the reaction in Equation 2.2:

$$\Delta S = S_{\text{albite}} - (S_{\text{jadeite}} + S_{\text{quartz}})$$
$$\Delta V = V_{\text{albite}} - (V_{\text{jadeite}} + V_{\text{quartz}})$$

S and V represent the molar entropy and molar volume of each phase, which can be looked up in tables of thermodynamic data for minerals (such as Robie *et al.*, 1979).

Equation 2.7 is called the **Clapeyron equation**, after is originator, an eminent 19th-century French railway engineer. It provides a means of estimating the gradient of a reaction boundary in a $P-T$ diagram from easily obtainable thermodynamic data. It is also very helpful in interpreting many features of phase-equilibrium diagrams.

To predict the slope of the phase boundary in Figure 2.2, one proceeds as follows. The book by Robie *et al.* (1979) gives the following molar entropy and molar volume data:

	S $J K^{-1} mol^{-1}$	V $10^{-6} m^3 mol^{-1}$
jadeite ($NaAlSi_2O_6$)	133.5	60.4
quartz (SiO_2)	41.5	22.7
albite ($NaAlSi_3O_8$)	207.4	100.1

Adding the reactants together:

$$\text{jadeite} + \text{quartz} \quad S = 175.0 \quad V = 83.1$$

Therefore ΔS, ΔV are $\quad +32.4 \quad +17.0$

Thus

$$\frac{dP}{dT} = \frac{32.4}{17.0} \frac{J\,mol^{-1}\,K^{-1}}{10^{-6}\,m^3\,mol^{-1}}$$
$$= 1.91 \times 10^6 \, J\,m^{-3}\,K^{-1}$$
$$= 1.91 \times 10^6 \, N\,m^{-2}\,K^{-1}$$
$$= 19.1 \times 10^5 \, Pa\,K^{-1}$$

The units ($10^5\,Pa\,K^{-1} = bar\,K^{-1}$) relate to the gradient of a line in $P-T$ space (Figure 2.2). The sign of the gradient is positive, consistent with Figure 2.2 (P rises with T), and the magnitude (19.1×10^5) agrees well with the value of about $20 \times 10^5\,Pa\,K^{-1}$ as measured from Figure 2.2.

The positive slope of Figure 2.2 thus reflects the observation that both ΔS and ΔV for this reaction are positve (or both negative, if we write the reaction the other way round). A negative slope would signify that ΔS and ΔV had opposite signs, as is the case for the andalusite–sillimanite reaction (Figure 2.1).

The most striking difference between Figures 1.3a, 2.1 and 2.2 on the one hand and Figure 2.3 on the other is that the latter has a curved reaction boundary, whereas the others are straight. The reason is that the volume change for such a reaction (and therefore dP/dT) is very pressure-sensitive:

$$\text{muscovite} = \text{sanidine} + \text{corundum} + \text{vapour}$$

$$\Delta V = V_{vapour} + V_{sanidine} + V_{corundum} - V_{muscovite}$$

At low pressures the volume of the 'vapour' phase (actually a super-critical fluid – see Box 2.2) is much greater than those of the solid phases, and therefore dominates the value of ΔV.

$$\Delta V \cong V_{vapour}$$

Because this term is large, the reaction boundary at low pressure has only a moderate slope. But the vapour, like any gas, is much more **compressible** than the solid phases. At higher pressures, V_{vapour} and ΔV will get progressively less, and the slope of the dehydration boundary will get correspondingly steeper. This general shape is a feature of all reactions involving the generation of a 'vapour' phase (see also Box 2.1, diagram A). Curved boundaries in $P–T$ diagrams always singify the involvement of a highly compressible phase, usually a gas (for example diagram B in Box 2.2).

2.5 Phase diagrams in $T–X$ space

2.5.1 Crystallization in systems with no solid solution

$P–T$ and $P_V–T$ diagrams make no provision for changes in the compositions of individual phases during reactions. Such changes are an important feature of igneous processes, and make it necessary to introduce another type of diagram in which the temperature of equilibrium is plotted as a function of phase composition ('X'). An example is shown in Figure 2.4, which shows the phase relations at atmospheric pressure for the binary system $CaMgSi_2O_6–CaAl_2Si_2O_8$. Because this system is relevant to igneous rocks (it includes simple analogues of basalt), the temperature range extends up far enough to include melting.

If the temperature is sufficiently high, it is possible to make a homogeneous melt containing the two components $CaMgSi_2O_6$ and

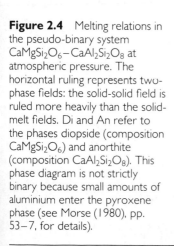

Figure 2.4 Melting relations in the pseudo-binary system $CaMgSi_2O_6–CaAl_2Si_2O_8$ at atmospheric pressure. The horizontal ruling represents two-phase fields: the solid-solid field is ruled more heavily than the solid-melt fields. Di and An refer to the phases diopside (composition $CaMgSi_2O_6$) and anorthite (composition $CaAl_2Si_2O_8$). This phase diagram is not strictly binary because small amounts of aluminium enter the pyroxene phase (see Morse (1980), pp. 53–7, for details).

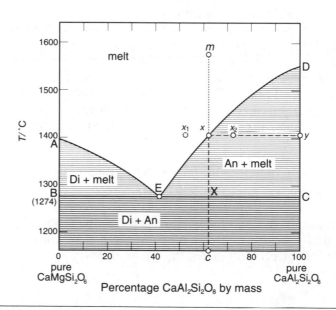

$CaAl_2Si_2O_8$ in any desired proportion. These compounds are said to be completely **miscible** in the melt phase. Consequently, in the field marked 'melt', only this single phase is stable. In the solid state, however, the two components exist as separate phases diopside (composition $CaMgSi_2O_6$) and anorthite (composition $CaAl_2Si_2O_8$): there is no stable homogeneous solid of intermediate composition. The area below 1274 °C is therefore a two-phase field.

The two areas ABE and ECD are also two-phase fields, each representing equilibrium between a melt and one of the crystalline phases. To see how, consider the line xy. This is an isothermal line at a temperature whose precise value is unimportant (in this case it is 1400 °C). We call this a **tie-line**, because it links together the compositions of two phases which can coexist stably at this temperature. x represents the only composition of the melt that can be in equilibrium with anorthite (composition y) at 1400 °C; it consists of 62% $CaAl_2Si_2O_8$ and 38% $CaMgSi_2O_6$. If the melt were more $CaMgSi_2O_6$-rich than this (composition x_1 for example), it would dissolve anorthite crystals and thereby increase the $CaAl_2Si_2O_8$ content until equilibrium was reached or until the anorthite present had all dissolved. If the liquid had the composition x_2, it would precipitate anorthite, so reducing the $CaAl_2Si_2O_8$ content.

As the line ED shows, the composition of melt that is in equilibrium with anorthite (An) depends on the temperature. The corresponding line A–E shows that the same is true of the melts that can coexist with diopside (Di). The curve AED, the locus of the melt compositions that can coexist in equilibrium with either diopside or anorthite at different temperatures, is called the **liquidus**. All states of the system lying above it in Figure 2.4 consist entirely of the melt phase. The one point common to both limbs of the liquidus is E, which therefore represents the unique combination of melt composition and temperature at which all three phases are simultaneously at equilibrium. This condition is called a **eutectic**.

In applying the Phase Rule to Figure 2.4, we must recognize that a $T-X$ diagram is no more than the end-view of more complex phase relations encountered in $P-T-X$ space (compare Figure 2.6a). By considering melting relations only at a single – in this case atmospheric – pressure, we are in fact artificially restricting the variance of each equilibrium. Any statements we make about variance in this diagram relate only to an apparent variance F' where:

$$F' = F - 1 \qquad (2.8)$$

One may write the Phase Rule in terms of F' as follows:

$$\phi + F = \phi + (F' + 1) = C + 2$$

Therefore $\qquad\qquad \phi + F' = C + 1 \qquad (2.9)$

37

This form of the Phase Rule, applicable to isobaric $T–X$ (and, incidentally, isothermal $P–X$) phase diagrams, is sometimes known as the **Condensed Phase Rule**.

Point E $\phi = 3$ (3 phases, Di + An + melt)
                   $C = 2$ (2 components)
          $3 + F' = 2 + 1$
Therefore $F' = 0$ an isobarically invariant equilibrium

The term isobarically invariant is jargon that reminds us that this equilibrium is invariant only so long as the pressure is held constant; if this constraint were to be relaxed, we would find that the eutectic was just one point of a univariant curve in $T–X–P$ space.

A eutectic is always an invariant point in any phase diagram. On the other hand, the one-phase 'melt' field has two degrees of freedom:

Point x_1 $\phi = 1$ (1 phase, melt)
                   $C = 2$ (2 components)
          $1 + F' = 2 + 1$
Therefore $F' = 2$ an isobarically divariant equilibrium

T and X have to be quantified to define the state of the system in this condition.

It would be natural to expect the fields ECD and ABE to be divariant as well, but the Phase Rule indicates otherwise. Consider the composition represented by point x_2. Neither the melt (at this temperature) nor anorthite can have this composition. The 'composition' x_2 only has meaning at 1400 °C when interpreted as the composition of a physical mixture of melt x and anorthite y. (The proportions of the two phases in this mixture can be worked out as explained in Box 2.3.) ECD and ABE are therefore two-phase fields. If $\phi = 2$ and $C = 2$, we cannot escape the conclusion that $F' = 1$. In other words, specifying temperature is sufficient to define the composition of all phases in equilibrium, or vice versa given that the pressure is already defined. Thus reaction-boundary lines (cf. Figures 1.3a and 2.1) are not the only manifestation of univariant equilibrium in phase diagrams; areas can also be univariant. Such fields arise in $T–X$ diagrams whenever two coexisting phases have different compositions. One can imagine them consisting of an infinity of horizontal tie-lines, as the horizontal ruling in Figure 2.4 is intended to symbolize.

Box 2.3

Tie-lines and the lever rule

In $T-X$, $X-X$ or $P-T-X$ diagrams, tie-lines link together two different compositions which can coexist in equilibrium under specific conditions. Any composition lying between the ends of a tie-line must therefore represent a mixture of the two phases. From the position of that point on the tie-line one can work out the relative proportions of the two phases in the mixture.

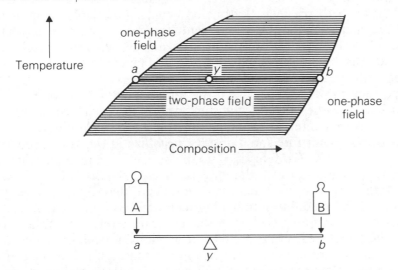

The upper diagram shows two phases of compositions a and b in equilibrium at a particular temperature. Point y in the two-phase field between them represents the composition of a physical mixture of phases a and b. If y is pushed leftwards until it coincides with a (the mixture has the same overall composition as phase a), it symbolizes a 'mixture' consisting of 100% of a and no b. If y coincides with b, the 'mixture' will consist entirely of b. For intermediate mixtures it can be shown that:

$$\text{percentage of phase } a \text{ in mixture} \propto \text{distance } y\text{–}b$$

$$\text{percentage of phase } b \text{ in mixture} \propto \text{distance } a\text{–}y$$

Therefore

$$\frac{\text{percentage of } a}{\text{percentage of } b} = \frac{\text{distance } y\text{–}b}{\text{distance } a\text{–}y}$$

This useful equation is known as the **Lever Rule**, owing to its similarity to the old-fashioned beam-balance (lower diagram), in which the weight of a body A is inversely proportional to the distance from the fulcrum $(a\text{–}y)$ at which it balances an opposing weight B:

$$\frac{\text{weight of A}}{\text{weight of B}} = \frac{y\text{–}b}{a\text{–}y}$$

Qualitatively, the closer the composition of a mixture plots (in composition space) to one of its constituents, the greater the percentage of that constituent in the mixture.

TX diagrams are important in igneous petrology because they allow one to follow the evolution of melt composition with advancing crystallization in experimental and natural magmatic systems (at constant pressure). Imagine a melt *m* cooling from some temperature above the liquids, say 1500 °C. At first there will be no change other than a fall in temperature: we can imagine point *m* falling vertically through the 'melt' field. Arrival at the liquidus (point *x*) signals the first appearance of solid anorthite, which here begins to crystallize in equilibrium with the melt. Extraction of anorthite depletes the melt a little bit in the $CaAl_2Si_2O_8$ component, causing a shift in composition to the left in Figure 2.4. But the maintenance of univariant equilibrium demands that the melt composition should change in conjunction with falling temperature. As crystallization advances, therefore, the melt composition moves steadily down the liquidus curve towards E.

On reaching the eutectic, the melt begins to crystallize diopside in addition to anorthite. At this juncture, the melt composition becomes fixed, because anorthite and diopside crystallize in the same proportion as the $CaAl_2Si_2O_8 : CaMgSi_2O_6$ ratio of the melt. The temperature also remains constant, because the eutectic is an invariant equilibrium (at least within the isobaric framework being considered): as long as three phases remain in equilibrium neither melt composition nor temperature can change. Continued cooling in this context merely means the loss of heat (the latent heat of crystallization of diopside and anorthite) from the system at constant temperature and the formation of crystals at the expense of melt. Eventually the melt becomes exhausted, and invariant (Di + An + melt) equilibrium gives way to univariant (Di + An) equilibrium, allowing the temperature to resume its downward progress. The total solid assemblage will obviously now consist of 62% anorthite and 38% diopside (*c* in Figure 2.4).

The eutectic therefore represents the lowest projection of the melt field, the composition and temperature of the last melt to survive during the cooling of the system. Progressive crystallization of any melt in this system (with the special exceptions of pure $CaMgSi_2O_6$ and pure $CaAl_2Si_2O_8$) will lead its composition ultimately to the eutectic. This illustrates an important general principle of petrology, that the evolving compositions of crystallizing magmas of all types tend to converge on one or two 'residual magma' compositions (a natural example being granite) at which the liquidus reaches its lowest temperature.

The eutectic also indicates the composition of the first melt to appear on heating any mixture of diopside and anorthite (Box 2.4).

Eutectics are common features in systems of this kind. They represent the general observation that mixtures of minerals (in other words, rocks) begin to melt at lower temperatures than any of the pure constituents (minerals) would on their own, just as a mixture of ice and salt conveniently melts at a lower temperature than ice alone. This principle is widely used in industry when a flux is added to enable a substance to melt at a lower temperature than it would in the pure state (e.g. in soldering).

2.5.2   Crystallization in systems with solid solution

Diopside and anorthite belong to different mineral groups having different crystal structures, and the tendency for either to incorporate the constituents of the other into its crystal structure is negligible. But within many mineral groups it is common to find that crystal composition can vary continuously between one end-member composition and another. One can think of one solid end-member dissolving in the other to form a homogeneous crystal of intermediate composition. This phenomenon of miscibility in the solid state is referred to as **solid solution**.

Figure 2.5 shows the crystallization behaviour of a familiar example of such solid solution series, plagioclase feldspar (a solid solution between albite, $NaAlSi_3O_8$, and anorthite, $CaAl_2Si_2O_8$). Only one solid phase appears in the diagram.

Prominent in the diagram is a leaf-shaped feature, bounded by two curves depicting coexisting crystal and melt compositions as a function of temperature. Thus line ab in Figure 2.5 is a tie-line linking melt composition a to the crystal composition b with which it is in equilibrium at that temperature. The curve through a, above which (in the 'melt'

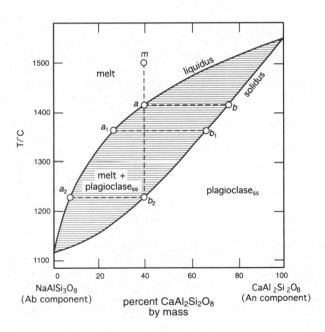

Figure 2.5 $T-X$ diagram showing melting relations in plagioclase feldspars at atmospheric pressure. Horizontal ruling represents a two-phase field. The subscript 'ss' denotes that plagioclase is a solid solution (see text).

Box 2.4

Partial melting I: melting in the laboratory

Phase diagrams provide many insights into the important process of rock melting. Consider first the progressive melting of a mixture of diopside and anorthite, for example the mixture of composition c in Figure 2.4.

The mixture will simply heat up until the temperature reaches 1274°C. At this point a melt of the eutectic composition E begins to form; this is the only melt composition that can be in equilibrium with diopside and anorthite, both of which are at this stage present in the solid mixture. Continued heating brings an increase in the proportion of melt at constant temperature, with no change of melt composition (invariant equilibrium) until the diopside disappears, having been entirely incorporated in the melt. (Anorthite has been dissolving too, but has not yet been used up.) At this stage the ratio of melt:anorthite is XC:EX. Univariant equilibrium (An + melt) now obtains, and with increasing temperature the proportion of melt continues to rise, its composition proceeding up the liquidus curve as more and more anorthite dissolves in it. At x, the melt has the same composition as the solid starting mixture, and here the last crystals of anorthite disappear. The system now enters the divariant melt field, where the temperature can continue rising without further change of state.

Similar principles govern the melting of solid-solution minerals. The olivine system shown in diagram A (analogous to Figure 2.5) provides an example relevant to basalt production by partial melting in the upper mantle (which consists chiefly of olivine). When olivine is heated up to the solidus (for example, point c_1), a small proportion of melt m_1 appears. The melt is much less magnesian than the olivine from which it is produced, a point of great petrological significance. Continued heating will cause the temperature to rise, the proportion of melt to increase, its composition to migrate up the liquidus curve towards m_2, and that of the remaining olivine crystals to migrate up the solidus towards c_2. The system would become completely molten at just over 1800°C (m_3). Thus a gap exists between the temperature at which olivine begins to melt and that at which it becomes completely liquid (as in Figure 2.5). This gap, called the **melting interval**, is a feature of all minerals that exhibit solid solution. The everyday notion of a melting point applies only to pure end-members, where the liquidus and solidus converge.

The complete melting of rocks like this only occurs in very unusual circumstances, like meteorite impacts. Generally magmas are produced by a process of **partial melting**, in which temperatures are sufficient to melt a fraction of the source material but not all of it. Both in the olivine diagram and in actual rocks, partial melting will generate a melt less magnesian (m_2) than the source material (c_1), leaving behind a refractory solid residue which is more magnesian (c_2) than

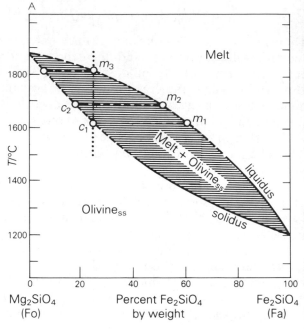

Melting relations in the olivine series at atmospheric pressure. This system has been determined experimentally only up to 1500°C, and for pure forsterite. Interpolated boundaries are shown dashed.

the source material prior to melting. The composition of both products, melt and residual solid, depends on the degree (percentage) of melting, and therefore on the temperature attained.

We must be careful not to assume that the temperatures shown in the olivine diagram are necessarily characteristic of the upper mantle. We have seen that although pure anorthite (Figure 2.4) remains solid up to 1553°C, a mixture of anorthite with diopside begins to melt below 1300°C. The same is true of mantle rocks, which consist not of olivine alone (though it is the chief constituent), but contain pyroxene and either garnet or spinel as well. Any such mixture will begin melting at temperatures below those at which the individual minerals would start to melt. Moreover every mineral contributes to the composition of the melt right from the start, just as both anorthite and diopside contribute to the eutectic melt in Figure 2.4. It is a common misconception to believe that minerals in a rock will melt one after the other. A partial melt should be seen as a solution in which all the solid phases of the source rock are partly soluble.

The influence of pressure (depth) on melting is considered in Box 2.5.

field) the system is entirely liquid, is the **liquidus**; and that through b, below which the system consists entirely of crystalline plagioclase, is called the **solidus**.

By applying the condensed phase rule to Figure 2.5, we find that both the 'melt' and 'plagioclase' fields are divariant: $\phi = 1$, $C = 2$ ($NaAlSi_3O_8 + CaAl_2Si_2O_8$), therefore $F' = 2$. The 'melt + plagioclase' field is another example of a univariant field (cf. Figure 2.4): $\phi = 2$, $C = 2$, therefore $F' = 1$. If equilibrium exists between melt and plagioclase then specifying T automatically defines the compositions of both phases. Conversely, knowing either phase composition defines the other composition and the temperature unambiguously. The line ab is one of an infinite series of such tie-lines traversing the two-phase field, as the horizontal ruling symbolizes.

Crystallization in this system leads to a series of continuously changing melt and solid compositions. Melt m, for example, will cool until it encounters the liquidus curve at a, where plagioclase b will begin to crystallize. Because b is more $CaAl_2Si_2O_8$-rich than a, its extraction will deplete the melt in $CaAl_2Si_2O_6$ and thereby enrich it in $NaAlSi_3O_8$; with continued cooling and crystallization the melt composition will move down the liquidus curve. The changing melt composition causes a corresponding evolution in the equilibrium composition of the plagioclase crystals. Not only will newly crystallized plagioclase be more albite-rich than b, but there will be a tendency for early formed crystals, by continuous exchange of Na, Ca, Al and Si, to re-equilibrate with later, more albite-rich fractions of the melt. Thus to maintain complete equilibrium as the melt evolves to point a_1 on the liquidus, all the crystals that have so far accumulated must adjust their composition to b_1 on the solidus. Such adjustment requires solid-state diffusion to and from the centre of each crystal, and is a slow process. Crystal growth during natural magma crystallization commonly proceeds too quickly to allow complete continuous equilibrium between crystals and magma: the surface layers readjust to changing melt composition, but the crystal interiors fall behind. The result is a compositional gradient between more anorthite-rich cores and more albite-rich margins of the plagioclase crystals, a phenomenon known among petrologists as **zoning** (Box 3.1). Zoning is seen in other mineral groups as well, notably pyroxene.

If crystallization proceeds slowly enough to permit continuous re-equilibration (an ideal situation known as **equilibrium crystallization**), the final liquid has the composition a_2 and the end-product is a mass of crystals all having the composition b_2, the same as the initial melt composition m. Imperfect re-equilibration between crystals and the evolving melt ties up a disproportionate amount of the anorthite component in the cores of early formed crystals owing to overgrowth or burial, and the melt can then evolve to compositions beyond a_2 before it runs out, creating late fractions of plagioclase more albitic than the original melt. This process, in which isolation of early formed solids

43

allows later melts to develop to more extreme compositions, is called **fractional crystallization**. Crystallization of natural melts in crustal magma chambers approximates closely to fractional crystallization, and contributes a lot to the chemical diversity of igneous rocks and magmas.

2.5.3 The solvus and exsolution

The final $T-X$ section to be examined (Figure 2.6b) shows the phase relations of the alkali feldspars in the presence of water vapour (at a

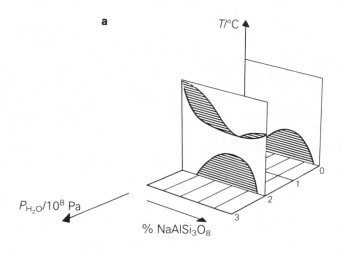

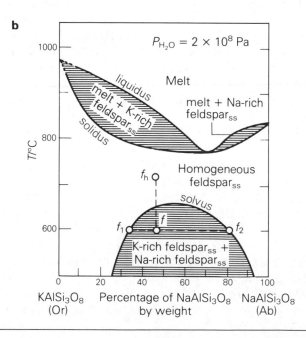

Figure 2.6 Melting and sub-solidus phase relations in the alkali feldspars (the system $KAlSi_3O_8-NaAlSi_3O_8$). The subscript 'ss' denotes solid solution. a. Perspective sketch of $P_{H_2O}-T-X$ space, showing the isobaric section at 2×10^8 Pa illustrated in b. b. Alkali feldspar phase relations at $P_{H_2O} = 2 \times 10^8$ Pa. Horizontal ruling represents two-phase fields.

Box 2.5

Partial melting II: melting in the Earth's mantle

The diagram below shows in simple terms how the solidus temperature of peridotite – the temperature at which it begins to melt – varies with depth in the mantle[1]. This is in essence a $P–T$ diagram, drawn 'upside-down' in terms of pressure in order to show temperature as a function of depth below the surface. The band between the solidus and liquidus curves, the 'melting interval', depicts the range of conditions under which partial melting (Box 2.4) of mantle peridotite can occur. These are the conditions necessary to produce basaltic magma.

The curve marked 'Oceanic geotherm' indicates how the ambient temperature in the upper mantle is believed to vary with depth beneath a typical sector of mid-ocean ridge. Note that the geotherm fails to reach the solidus at any depth. Why then should melting occur at all in the upper mantle?

To see why, we must recognize that the mantle is not a static body. Because the interior of the Earth is hot, the solid mantle undergoes continuous (though very slow) convective motion, with mushroom-like 'plumes' of buoyant hotter material ascending from below (e.g. beneath Hawaii), and dense colder material sinking down (e.g. cold oceanic lithosphere at subduction zones). In a convecting mantle, melting may arise purely as a consequence of upward motion. In diagram a, the solid peridotite at point X, for instance, can penetrate the solidus and begin to melt simply by migrating upward to lower pressures along the path X–Y. This process, known as **decompression melting**, is the primary cause of magma generation at mid-ocean ridges: plate forces continuously pull the lithospheric plates apart and thereby permit passive upwelling (diagram b) and melting of the underlying asthenosphere. No rise in temperature is required; indeed the ascending material cools slightly as a result of the work it has to do in expanding (Box 1.1).

Mantle plumes, on the other hand, are sites of buoyant upwelling of deeper material that may be 150–300 °C hotter than the surrounding upper mantle. Melting in a plume is the combined effect of an elevated geotherm and decompression resulting from upwelling. As many plumes are located in intra-plate settings, the presence of thick cool lithosphere confines melting to deeper levels than beneath a mid-ocean ridge.

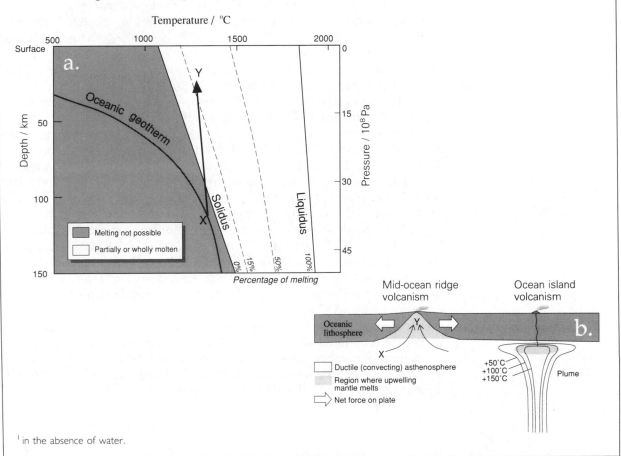

pressure of $2 \times 10^8 \, \text{Pa} = 2 \, \text{kbars}$). This diagram can be visualized as a cross-section of $P_{H_2O}-T-X$ space. coinciding with the plane in which $P_{H_2O} = 2 \times 10^8 \, \text{Pa}$ (Figure 2.6a). One can speak of the diagram as an **isobaric T–X section** of phase relations in $P_{H_2O}-T-X$ space.

The diagram shows the liquidus and solidus of the alkali feldspar series, which differ from those in Figure 2.5 only in that they fall to a minimum melting point in the middle of the series rather than at one end. As a result we get two leaf-shaped fields instead of one. But interest is mainly in what happens in the **sub-solidus region**. The 'homogeneous feldspar$_{ss}$' field immediately below the solidus means that here the end-members are completely miscible in the solid state: they form a complete solid solution in which any composition can exist as a single homogeneous phase. But at lower temperatures things get a little more complicated.

Beneath a boundary called the **solvus** a 'miscibility gap' appears. At these temperatures (for example 600 °C) the albite crystal structure is less tolerant of the $KAlSi_3O_8$ component (partly because of the large size of the potassium atom), and at a $KAlSi_3O_8$ content of about 20% (f_2) becomes **saturated** with it. Any additional $KAlSi_3O_8$ present is forced to exist as a separate $KAlSi_3O_8$-rich feldspar phase, whose composition can be found by extending a tie-line to the left-hand **limb** of the solvus, cutting it at f_1 (about 65% $KAlSi_3O_8$). This potassium feldspar is itself saturated with $NaAlSi_3O_8$.

A homogeneous alkali-feldspar solid solution such as f_h ceases to be stable as it cools through the solvus. At point f, for example, it is well inside the two-phase region. Such a point represents, at equilibrium, a mixture of two phases. The feldspar therefore breaks down, or **exsolves**, into two separate phases f_1 and f_2. But solid-state diffusion is too slow to allow a cooling feldspar to sort itself out into two separate crystals. The usual result of exsolution is a series of thin, lamellar domains of one phase enclosed within a host crystal of the other. The lever rule (Box 2.3) tells us that in the present example f_1 will be more abundant than f_2, and the cooling of crystal f_h will therefore produce a host crystal of composition f_1 containing **exsolution lamellae** of phase f_2. Such structures (illustrated in plates 80–1 in MacKenzie *et al.*, 1982) are characteristic of alkali feldspars and are called **perthites**. Similar structures (not given this name) are developed in some pyroxenes, owing to a similar miscibility gap between diopside ($CaMgSi_2O_6$) and enstatite ($Mg_2Si_2O_6$).

Application of the condensed phase rule (because P_{H_2O} is constant) to Figure 2.6b indicates that the sub-solvus region is a **univariant area**. In favourable circumstances, coexisting alkali feldspar (or pyroxene) compositions can be used to estimate the temperature of equilibration, a fact that has practical application in **geothermometry**.

Box 2.6

Reaction points and incongruent melting

Every geology student knows that olivine and quartz are incompatible, and to not coexist stably in nature. (In fact, this is true only of magnesium-rich forsteritic olivines. Fayalite is quite a common mineral in granites and quartz syenites.) How is this incompatibility expressed in a phase diagram?

The relevant part of the system $Mg_2SiO_4-SiO_2$ (omitting complications at the SiO_2-rich end) is shown for atmospheric pressure below. In many respects it is similar to Figure 2.4. The difference is that between Mg_2SiO_4 and SiO_2 lies the composition of the pyroxene enstatite, $Mg_2Si_2O_6$. Consider the crystallization of melt composition m_1. On reaching the liquidus it will begin to crystallize olivine, whereupon further cooling and crystallization will lead the melt composition down the liquidus curve. On reaching R, the melt composition has become too SiO_2-rich (more so than enstatite) to coexist stably with olivine, which therefore *reacts* with the SiO_2 in the melt to form crystals of enstatite:

$$Mg_2SiO_4 + SiO_2 \rightarrow Mg_2Si_2O_6 \quad (B2.5)$$
$$\text{olivine} \quad \text{melt} \quad \text{pyroxene}$$

(This symbolism does not mean that the melt consists of SiO_2 alone. Other components are present, but this reaction involves only the SiO_2 component.)

At R, the three phases are at equilibrium. Using the Condensed Phase Rule, it is clear that R is an invariant

point like E. It is called a **reaction point**. Temperature and melt composition remain constant as the reaction proceeds (from left to right in Equation B2.5), until one or other phase is exhausted. In this case (beginning with m_1) the melt is used up first, and the final result is a mixture of olivine and enstatite: the melt never makes it to the eutectic. If on the other hand the initial melt had the composition m_2, more siliceous than enstatite, the reaction at R would transform all of the olivine into enstatite, with some melt left over. The disappearance of olivine releases the system from invariant equilibrium R, and the melt can proceed down the remaining liquidus curve, crystallizing enstatite directly until the eutectic is reached. The final result is a mixture of enstatite and silica (the high-temperature polymorph cristobalite). The proportions in the final mixture can be worked out by applying the Lever Rule to m_2.

During melting, this reaction relationship manifests itself as a phenomenon called **incongruent melting**. Pure enstatite, when heated, does not melt like olivine or anorthite but decomposes at 1557 °C to form olivine (less SiO_2-rich) and melt (more SiO_2-rich than itself), the reaction B2.5 run in reverse. The system is held in invariant three-phase equilibrium until the enstatite has been exhausted, then continues melting by progressive incorporation of olivine into the melt (Figure 2.4).

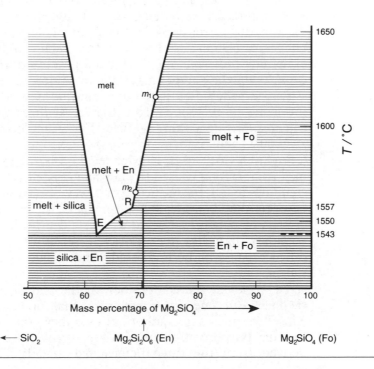

2.6 Ternary phase diagrams

Merely to represent the range of compositions possible in a three-component system requires the use of the two dimensions of a piece of paper ('Ternary Diagrams' in Appendix B). If we wish to represent comprehensively the phase relations of such a system over a range of temperature (in a form analogous to Figure 2.4), we must build a three-dimensional model. But, as a medium for transmitting phase equilibrium data around the world, such models would be less than convenient, so petrologists have devised a number of ways of condensing their content into two-dimensional form. Some examples are shown in Figure 2.7.

The base of any ternary phase diagram is an equilateral triangle, in which any composition in a ternary system can be plotted (Appendix B). The temperature axis of the hypothetical model is constructed perpendicular to the plane of this triangle. The gross features of the model can be presented in the form of a perspective sketch (Figure 2.7a) showing the topography of the liquidus surfaces (analogous to the liquidus curves in Figures 2.4–6), which meet in a V-shaped low-temperature trough running out of the binary eutectic in the system $CaMgSi_2O_6-CaAl_2Si_2O_8$. The phase relations in this and the companion binary systems (Di–Ab and An–Ab) can be indicated on the vertical faces of the 'model'.

Such sketches fail to represent the quantitative detail which the experimenter has laboured hard to obtain. A more exact way to represent the liquidus surfaces is in the form of a map contoured at specific temperature intervals (Figure 2.7b). Such a diagram is invaluable for examining the evolution of melt composition during crystallization, and considering the parallel magmatic evolution in real igneous rocks. The V-shaped valley divides the diagram into two fields, each labelled with the name of the solid phase that crystallizes first from melts whose compositions lie within that field. For example, a melt of composition a at 1300 °C will initially crystallize diopside. A line drawn from the $CaMgSi_2O_6$ apex to a, if extended beyond a, indicates the changes in melt composition caused by diopside crystallization. If the temperature continues to fall, the melt composition will eventually reach the boundary between the diopside and plagioclase$_{ss}$ fields (at point b), and here crystals of plagioclase begin to crystallize together with diopside. The boundary indicates the restricted series of melt compositions that can coexist with both diopside and plagioclase at the temperatures shown.

To work out the composition of the plagioclase requires the use of tie-lines. But one must remember that tie-lines are isothermal lines (since two phases in equilibrium must have the same temperature), and for this purpose it is appropriate to use a second type of diagram derived from the three-dimensional model. This is the **isothermal**

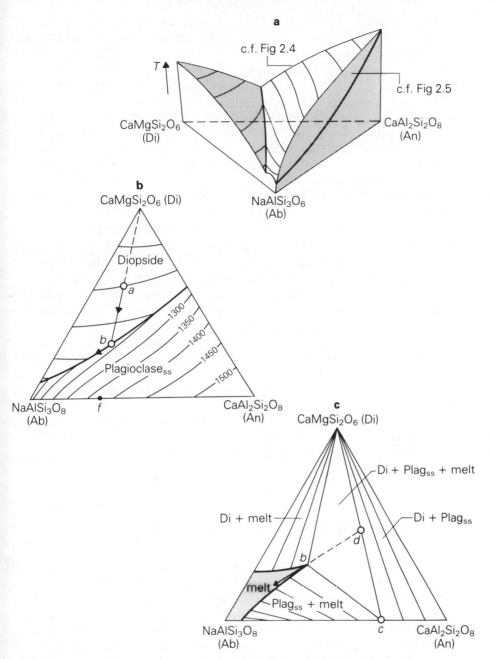

Figure 2.7 Ways of representing crystallization in the pseudo-ternary system CaMgSi$_2$O$_6$–NaAlSi$_3$O$_8$–CaAl$_2$Si$_2$O$_8$. a. A perspective sketch of the liquids surface in three dimensions. Elevation and contours represent temperature. b. A plan view of the liquidus surface, with topography shown by temperature contours (graduated in °C). c. An isothermal section at 1220 °C. Tie-lines across the two-phase fields show the composition of coexisting phases. The arrow (tangent to the cotectic curve in Figure 2.7b) indicates the direction in which the melt composition b will evolve with further crystallization. This direction is controlled by the proportion in which diopside and plagioclase (c) crystallize (point d).

49

section shown in Figure 2.7c. In principle an isothermal section can be drawn for any temperature for which phase equilibrium data are available: this one relates to 1220 °C, the temperature of the liquidus at point b in Figure 2.7b. The area where the liquidus lies below the temperature of the section (stippled in Figure 2.7c) is a one-phase field where only melt is stable. The rest of the diagram can be regarded as the result of slicing the top off the solid model at 1220 °C, revealing three two-phase fields, each traversed by a family of tie-lines (summarizing the results of phase equilibrium experiments at 1220 °C). The composition of the plagioclase$_{ss}$ in equilibrium with the melt b at this temperature can be read from the diagram by following the tie-line from b to the $NaAlSi_3O_8$–$CaAl_2Si_2O_8$ edge of the diagram (point c).

Tie-line b–c forms one boundary of a **three-phase field** representing equilibrium between melt b, plagioclase c and diopside (composition $CaMgSi_2O_6$) at this temperature. Any point lying within this field signifies a physical mixture of these coexisting phases, the proportions of which could be worked out using the Lever Rule. Because all possible mixtures of diopside and plagioclase c lie to the right of melt b (along the line Di–c), the crystallization of these two minerals with cooling causes the melt composition to migrate leftwards, along the boundary – the **cotectic** – shown in Figure 2.7b. This direction is indicated by the arrow in Figure 2.7c. Point d, collinear with the arrow, indicates the proportion in which diopside and plagioclase (c) crystallize from the melt b (Exercise 5).

More detailed interpretation of such diagrams lies beyond the scope of this book. The interested reader should look at the books by Morse (1980) and Ernst (1976), both of which devote several pages of this particular system.

2.7 Review

The great diversity of reactions and assemblages recorded in natural igneous and metamorphic rocks provides many avenues for investigating the conditions under which the rocks were formed. We have seen that, to analyse what such mineral assemblages mean in terms of pressure and temperature of formation, we can draw on two sorts of published experimental information. The primary source is the literature of **experimental petrology**, in which one can usually track down a number of phase diagrams relevant to the assemblages in a particular rock suite. Such diagrams, derived from well-established laboratory procedures (Box 2.1) which are familiar to most petrologists, present phase equilibrium data in an easily understood form. But many diagrams of this kind refer to experiments on simple laboratory analogues rather than on the rocks themselves. The melts in Figure 2.7, for example, fall short of having true basaltic compositions, owing to the absence, among other things, of the important element iron. (One of the many consequences of this defect is that equilibria in Figure 2.7b are shifted to higher temperatures than would be found in a real iron-bearing basalt.) Thus diagrams like Figure 2.7, though invaluable for analysing principles of phase equilibrium, do not reflect in quantitative detail the behaviour of more complex natural magmas and rocks. In some circumstances it can be helpful to carry out experiments on natural rock powders or comparable synthetic preparations, but the results cannot be directly displayed in simple phase diagrams and have less general application.

There are also useful applications in petrology for **thermodynamic** data (molar enthalpies, entropies and volumes). Using the Clapeyron equation and Le Chatelier's principle, we can predict certain features of phase diagrams without recourse to petrological experiment. Molar enthalpies and entropies of pure minerals are measured primarily by a completely different technique called **calorimetry**, involving the very accurate measurement of the heat evolved when a mineral is formed from its constituent elements or oxides. Such methods and data have, until recently, been less familiar to geologists, and their successful application in solving petrological problems requires a command of thermodynamic theory beyond the scope of this book. Thermodynamics has, however, become one of the most versatile tools of metamorphic petrologists, enabling them to apply experimental data from simple synthetic systems to complex natural assemblages, and one can expect fruitful developments in this area in future years. The interested reader can pursue this subject in the introductory book by Wood and Fraser (1976).

2.8 Bibliography

Brown, G. C., Hawkesworth, C. J. and Wilson, R. C. L. (1992) *Understanding the Earth – a new synthesis*, Cambridge University Press, Cambridge. Chapters 5 and 6.

Cox, K. G., Bell, J. D. and Pankhurst, R. J. (1979) *The interpretation of igneous rocks*, George Allen and Unwin, London.

Ehlers, E. G. (1972) *The interpretation of geological phase diagrams*, W. H. Freeman, San Francisco.

Ernst, W. G. (1976) *Petrologic phase equilibria*, W. H. Freeman, San Francisco.

Fletcher, P. (1993) *Chemical thermodynamics for Earth scientists*, Longman, Harlow.

Holloway, J. R. and Wood, B. J. (1988) *Simulating the Earth – experimental geochemistry*, Unwin Hyman, Boston.

MacKenzie, W. S., Donaldson, C. H. and Guilford, C. (1982) *Atlas of igneous rocks and their textures*, Longman, London.

Morse, S. A. (1980) *Basalts and phase diagrams*, Springer-Verlag, New York.

Nordstrom, P. K. and Munoz, J. L. (1986) *Geochemical thermodynamics*, Blackwell, Oxford.

Yardley, B. W. D. (1989) *An introduction to metamorphic petrology*, Longman, Harlow.

Wood, B. J. and Fraser, D. G. (1976) *Elementary thermodynamics for geologists*, Clarendon Press, Oxford.

2.8.1 Sources of thermodynamic data for minerals

Berman, R. G. (1988) Internally consistent thermodynamic data for minerals in the system $Na_2O-K_2O-CaO-MgO-FeO-Fe_2O_3-Al_2O_3-SiO_2-CO_2$, representation, estimation and high temperature extrapolation. *Contr. Mineral. Petrol.*, **89**, 168–283.

Powell, R. (1978) *Equilibrium thermodynamics in petrology: an introduction*, Appendix A, Harper and Row, London.

Robie, R. A., Hemingway, B. S. and Fisher, J. R. (1979) *Thermodynamic properties of minerals and related substances at 298.15 K and 1 bar (10^5 Pa) pressure and at higher temperatures*, US Geological Survey, Bulletin 1452. (A standard source of thermodynamic data for minerals.)

2.9 Exercises

1 Applying the Phase Rule to the reaction shown in Equation 2.4, discuss the variance of points X and Y in diagram a, Box 2.1. Identify the degrees of freedom operating at point X.

2 Why does ice floating on water tell us that the melting temperature of ice will be depressed at high pressures?

3 At atmospheric pressure (10^5 Pa), the following reaction occurs at 520 °C:

$$Ca_3Al_2Si_3O_{12} + SiO_2 \rightarrow CaAl_2Si_2O_8 + 2CaSiO_3$$

grossular   quartz   anorthite   wollastonite

(a garnet)   (plagioclase)

Use the data below to plot a correctly labelled *P–T* diagram for pressures up to 10^9 Pa.

	Entropy S $J\,K^{-1}\,mol^{-1}$	Volume V $10^{-6}\,m\,mol^{-1}$
grossular ($Ca_3Al_2Si_3O_{12}$)	241.4	125.3
quartz (SiO_2)	41.5	22.7
anorthite ($CaAl_2Si_2O_8$)	202.7	100.8
wollastonite ($CaSiO_3$)	82.0	39.9

4 Refer to Figure 2.5. Calculate the relative proportions of melt and crystals produced by cooling a melt of composition m to (a) 1400 °C, (b) 1300 °C, and (c) 1230 °C. What are the compositions of melt and plagioclase at these temperatures? (Assume that equilibrium is maintained throughout.)

5 Refer to Figure 2.7 and its caption. In what proportions must diopside and plagioclase crystallize from melt b to drive its composition along the cotectic curve (the arrow in Fig. 2.7c)?

Calculate the compositions and proportions of the phases present in a solid mixture of composition a (Fig. 2.7b). What would be the equilibrium assemblage for this mixture at 1220 °C? What would be the compositions and relative proportions of the phases present?

6 The figure below shows the ternary eutectic in the system $CaMgSi_2O_6$ (Di)–Mg_2SiO_4 (Fo)–$Mg_2Si_2O_6$ (En) determined from experiments carried out at a pressure of 20×10^8 Pa. To what depth in the mantle does this pressure correspond? What is the composition (expressed as $Di_xFo_yEn_z$ where x, y and z are per-

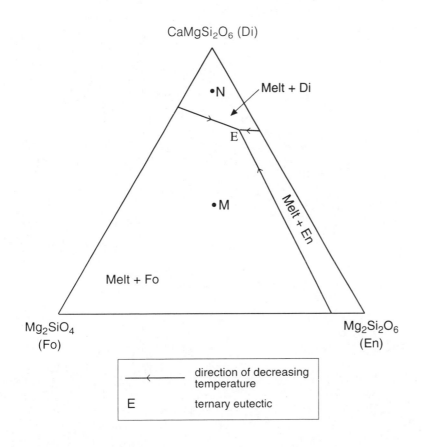

centages) of the first melt to form as (a) the mixture M and (b) the mixture N is heated through the solidus?

3
Kinetics of geological processes

In reading Chapter 2, it is easy to fall into the trap of supposing that reactions between minerals are always rapid enough to reach chemical equilibrium during the time available, however short that may be. A little thought however will indicate that this cannot be so. We have seen (Figure 1.3) that aragonite is a metastable mineral at atmospheric pressure, and its survival in some outcrops of metamorphic rocks crystallized at high pressure is a sign of **disequilibrium**: the rate of inversion to calcite has been too slow for the reaction to be completed before the process of uplift and erosion brought the rock to the surface. The same reasoning applies to occurrences in surface outcrops of the mineral sillimanite, which is stable only at elevated temperatures (Figure 2.1). Examination of igneous or metamorphic rocks in thin section often brings to light petrographic evidence of disequilibrium, in such forms as mineral zoning and corona structures (Box 3.1). From these examples it is clear that the rate of progress of geochemical reactions, and the way it responds to different conditions, are factors we cannot ignore.

The measurement and analysis of chemical reaction rates is called **chemical kinetics**, a science whose simpler geological applications will be considered in the present chapter. Chemical kinetic theory provides the basis of understanding how (and why) reaction rates depend on temperature, a matter of fundamental geological importance. It also provides the algebra upon which radiometric (isotopic) dating methods are based (Box 3.2).

3.1 Defining the rate of a reaction

It is easy to accept that some reactions proceed faster than others, but less easy to see how such differences can be expressed quantitatively. What precisely do we mean by the **rate** of a reaction?

Box 3.1

Disequilibrium textures

Mineral reactions that are unable to proceed to completion leave a rock in a state of chemical disequilibrum which, on the scale of a thin section, may be indicated by a variety of disequilibrium textures.

Coronas

Diagram A shows a possible cooling path for a gabbro that crystallizes at a deep level in the crust. The principal minerals crystallizing from the melt will be orthopyroxene (opx), aguite (aug) and plagioclase (plag). The orthopyroxene and plagioclase begin reacting with each other as the temperature falls below the reaction boundary shown, forming a new high-pressure (garnet-bearing) metamorphic assemblage. The reaction takes place only at grain boundaries where the two reactant phases are in mutual contact, and the reaction products (garnet and diopside) accumulate in concentric zones along the same boundaries, producing a **corona** (diagram B). The texture suggests the reaction slowed to a halt before it could use up all of the orthopyroxene or plagioclase. A similar corona, though not involving garnet, is figured by MacKenzie *et al.* (1982, plate 92).

Reaction rims

Similar textures called **reaction rims** arise in all sorts of igneous rocks, due to reaction between early-formed crystals with a later, more evolved melt. Olivine phenocrysts may become mantled by orthopyroxene owing to the reaction with residual liquid (Box 2.6), or pyroxene may be rimmed by amphibole owing to reaction with the dissolved water content of the magma (MacKenzie *et al.*, 1982: plate 92).

Zoning

In a rock that achieved complete chemical equilibrium between its phases at a given temperature, all minerals would be homogeneous in composition. Igneous and metamorphic minerals are, however, quite commonly zoned. Zoning indicates that intra-crystalline diffusion has failed to keep pace with changing circumstances. Zoning in igneous minerals (MacKenzie *et al.*, 1982: plates 96–102) often reflects chemical evolution of the melt (perhaps only on a very local scale) with which only the surface layer of the growing crystal has maintained equilibrium (diagram C). Zoning in igneous and metamorphic rocks may also be a response to changing physical conditions (*P*, *T*, etc.).

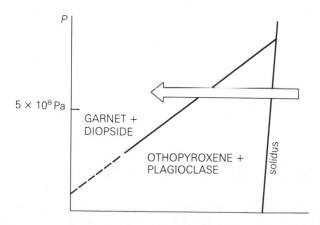

A. *P–T* diagram showing crystallization pathway of gabbro at a deep level in the crust.

Exsolution

Perthites (Mackenzie *et al.*, 1982: plates 80 and 81) and similar textures in pyroxenes (plate 82) represent the solid-state decomposition of a homogeneous crystal into two immiscible phases (Figure 2.6). Intra-crystalline exsolution lamellae have a very large area of interface with the host crystal. The mismatch of structure across this interface generates a large positive **interfacial energy**, a situation undoubtedly less stable than equilibrium segregation into separate crystals. The persistence of lamellae indicates that diffusion through the crystal was too slow to bring this about.

An example of how exsolution behaviour can be used to measure cooling rates is given in Box 3.5.

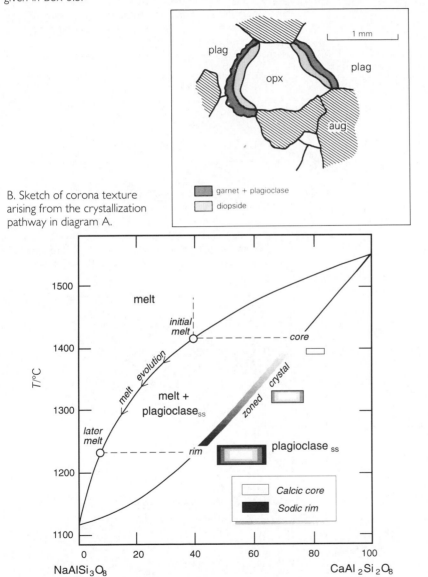

B. Sketch of corona texture arising from the crystallization pathway in diagram A.

C. Formation of a zoned plagioclase crystal as a result of the compositional evolution of the melt.

Box 3.2

Kinetics of radioactive decay I: the Rb–Sr system

The radioactive isotope of rubidium, ^{87}Rb (Box 10.1), decays to the strontium isotope ^{87}Sr. The kinetics of this nuclear reaction can be treated in the same way as a chemical reaction:

$$^{87}\text{Rb} \rightarrow \ ^{87}\text{Sr} \ + \beta^- + \bar{v}$$

'parent'   'daughter'
isotope   isotope

where the β-particle (β^-) and the anti-neutrino ($\bar{v}$) are released by the reaction.

The decay rate of any radioisotope is proportional to the number of the radioisotope nuclei present in the sample (N) at the moment in question. This can be written as a rate equation:

$$-\frac{dN}{dt} = \lambda N$$

Because there is only one concentration term (N) on the right-hand side (unlike Equation 3.3), this is called a **first-order reaction**. λ is the rate constant analogous to k in Equation 3.3, but in this context it is called the **decay constant**.

The rate equation can be **integrated** to show how N varies with time:

$$N = N_0 e^{-\lambda t}$$

where N_0 is the number of radioisotope nuclei initially present (when $t = 0$). The decay profile for a short-lived

isotope is shown in diagram A. If we transform both sides into natural logarithms and rearrange things a bit, we get:

$$\ln\left(\frac{N_0}{N}\right) = \lambda t$$

This equation is more useful because it is **linear** in relation to time (diagram B). For radioisotopes that decay rapidly, the decay constant can be determined by measuring the gradient of this graph (see Exercise 1).

The **half-life** $t_{\frac{1}{2}}$ of a radioisotope is the time it takes for N to decay to half of its original value ($N = \frac{1}{2}N_0$). Thus

$$\ln(2/1) = \lambda t_{\frac{1}{2}}$$

therefore $t_{\frac{1}{2}} = 0.6931/\lambda$

Because the decay of ^{87}Rb is very slow, the numerical value of λ is extremely small: 1.42×10^{-11} year^{-1}. During one year only about 14 out of every million million ^{87}Rb nuclei are likely to decay. The ^{87}Rb remaining in the Earth today has survived from element-forming processes (Chapter 10) that took place before the solar system was 4.6 billion years ago.

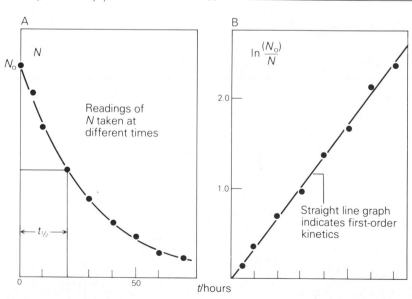

A — Readings of N taken at different times

B — Straight line graph indicates first-order kinetics

t/hours

Consider a simple chemical reaction, for example that between nitric oxide (NO) and ozone (a form of oxygen, O_3), two gaseous pollutants that occur in the troposphere as a result of the burning of fossil fuels. These **reactants** react with each other in equal molecular proportions to form the **products** nitrogen *di*oxide (NO_2) and ordinary oxygen (O_2), which are also gases:

$$NO + O_3 \rightarrow NO_2 + O_2 \tag{3.1}$$
$$\text{gas} \quad \text{gas} \quad \text{gas} \quad \text{gas}$$

Imagine an experiment in which NO and O_3 are reacted together in a sealed vessel equipped with sensors that monitor the changing concentrations of NO, O_3, NO_2 and O_2 as the reaction progresses. (How these sensors work need not concern us.) The reaction consumes NO and O_3, whose concentrations (c_{NO} and c_{O_3}, each expressed in $mol\,dm^{-3}$) therefore decrease with time as shown in Figure 3.1. The concentrations of the products increase correspondingly as the reaction proceeds. The rate of the reaction at any stage is the **gradient** of the right-hand graph in Figure 3.1 at the moment concerned. Borrowing the symbolism of calculus (Appendix B):

$$\text{rate} = \frac{dc_{NO_2}}{dt} = - \frac{dc_{NO}}{dt} \tag{3.2}$$

Because one mole of NO_2 is produced for every mole of NO that is consumed, the gradients of the left- and right-hand graphs in Figure 3.1 differ only in their sign (negative and positive respectively).

3.1.1 Rate equation

If we repeated the experiment with double the concentration of NO present (c_{O_3} being initially the same as before), we would find that the

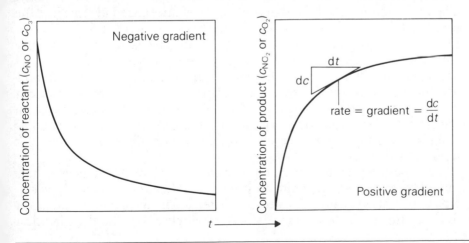

Figure 3.1 Composition–time curves for reaction 3.1.

initial rate is doubled. If we double the initial concentrations of both NO and O_3, we find the rate is quadrupled. This suggests that the rate is related to reactant concentrations:

$$\text{rate} = -\frac{dc_{NO}}{dt} = k \cdot c_{NO} \cdot c_{O_3} \qquad (3.3)$$

This equation is called the **rate equation** for this reaction. Because it contains two concentration terms (c_{NO} and c_{O_3}), the reaction is said to have second-order kinetics. The constant k, whose numerical value is specific to this reaction (and to the temperature at which the experiment is run), is called the **rate constant**. The equation predicts that as the reactants are used up the rate will decline, which is consistent with the flattening out of the slopes in Figure 3.1.

The process of radioactive decay can be analysed in a similar manner (Box 3.2).

3.1.2 Heterogeneous reactions

Reactions like 3.1 that take place within a single phase (in this case a gas mixture) are called **homogeneous reactions**. Nearly all reactions of geological significance on the other hand are **heterogeneous reactions**, involving two or more phases (minerals). Because they require the migration of components across the **interface** dividing one phase from another, the formulation of rate equations for heterogeneous rections is far more complicated than for homogeneous reactions.

The most obvious consequence of involving two phases in a reaction is that the surface area of their interface becomes a variable in the rate equation. Surface area is determined chiefly by particle size. The surface area of a cube 1 cm across is $6\,cm^2$ (six sides each of $1\,cm^2$ area). Cutting the cube in half in each direction produces eight cubes, each 0.5 cm across and each having a surface area of $6 \times (0.5)^2 = 1.5\,cm^2$. The total volume of all the cubes together is unchanged ($1\,cm^3$) but the total surface area has increased from $6\,cm^2$ to $8 \times 1.5 = 12\,cm^2$. Dividing the original cube into 1000 cubelets each of 0.1 cm size would increase the total area to $60\,cm^2$, while reducing to particle sizes equivalent to silt and clay sediments would increase their surface area to 3000 and $60\,000\,cm^2$ respectively. Particle or crystal size, because it determines the area of contact between phases, has a profound effect on the rate of a heterogeneous reaction. This is why diesel fuel injected as a fine spray into an engine reacts explosively with air, whereas the bulk liquid burns much more slowly.

The condition of the interface is also a very important factor. The rate of inversion of aragonite to calcite, for example, is greatly accelerated by the presence of traces of water along the grain boundaries. Surface chemistry has many important applications in the chemical

Box 3.3

Kinetics of radioactive decay II: the U–Pb system

Each of the naturally occurring isotopes of uranium (^{235}U and ^{238}U) decays through a complex series of intermediate radioactive nuclides to an isotope of lead (Pb). This is illustrated for ^{235}U in the figure. These decay chains form the basis of U–Pb radiometric dating. In spite of its complexity, the overall reaction conforms to first-order kinetics, because the first step in the process (to ^{231}Th, a radioactive isotope of thorium) happens to be the slowest. The kinetic complexities of the subsequent branching decay series are immaterial because the whole process is controlled by this one **rate-determining step**. just as the flow of water out of a hose can be controlled by constriction of the tap feeding it. The kinetics of complex chemical reactions may also be controlled by a slow, rate-determining step.

It is relevant to note that, for every uranium nucleus that decays to lead within the Earth, 7 or 8 alpha particles are released. By capturing electrons, those particles become ^{4}He atoms which form the bulk of the helium flux escaping from the Earth's interior.

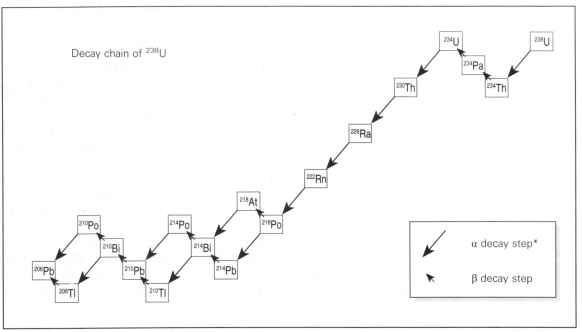

Decay chain of ^{238}U

$\nearrow$ α decay step*

$\nwarrow$ β decay step

*See Box 10.1

industry and in mineral processing (for example, the use of a frothing agent to optimize the separation of ore minerals by flotation).

Mechanical factors come into play as well. When a solid dissolves, the aqueous phase surrounding it becomes locally saturated, and impedes further solution until diffusion has distributed the dissolved species more evenly. Dissolution of sugar in coffee can therefore be accelerated by the use of a teaspoon to promote homogenization, and natural forms of agitation can be correspondingly effective in the marine environment. Experiments show that the rate at which calcite dissolves in water can be represented like this:

$$\text{Rate} = kA\alpha^{\frac{1}{3}}\{K^0 - (c_{Ca^{2+}})^{\frac{1}{2}}(c_{CO_3^-})^{\frac{1}{2}}\} \tag{3.4}$$

The c terms refer to concentrations in solution, K^0 and k are constants, A is the total surface area of the calcite phase present, and α is the experimental stirring rate (which appears as the cube root for reasons that are not clear). No doubt the effect of natural wave-agitation is still more complicated. This equation illustrates how rapidly the complexities can multiply when even the simplest heterogeneous reactions are studied kinetically.

3.2 Temperature-dependence of reaction rate

Everyday experience tells us that chemical reactions, whether homogeneous or heterogeneous, speed up as the temperature is raised. Epoxy adhesives cure more quickly in a warm oven. The very fact that we use refrigerators indicates that biochemical reactions slow down at lower temperatures. Quantitatively the temperature effect which these examples illustrate is quite pronounced: many laboratory reactions roughly double their reaction rates when the temperature is raised by just 10 °C (Exercise 2 at the end of this chapter). The temperature-dependence of reaction rates is particularly significant for geological processes, whose environments can vary in temperature over many hundreds of degrss.

Most reactions vary with temperature in the manner shown in Figure 3.2a. The 19th-century Swedish physical chemist S. A. Arrhenius showed that this behaviour could be represented algebraically by expressing the rate constant for a reaction in terms of an exponential equation:

$$k = Ae^{-E_a/RT} \tag{3.5}$$

This has become known as the **Arrhenius equation** and it has a number of important applications in geology. R is the **gas constant** ($8.314\,\text{J}\,\text{mol}^{-1}\,\text{K}^{-1}$) and T is the temperature *in kelvins*. A and E_a are constants for the reaction to which the rate constant refers (they vary from one reaction to another). A is called the pre-exponential factor, and it has the same units as the rate constant (which depend on the order of the reaction concerned). The constant E_a is called the **activation energy** of the reaction.

In Chapter 1 it was shown that physical and chemical processes often encounter an energy 'hurdle' which impedes progress from the initial, high-energy (less stable) state to the lower-energy configuration in

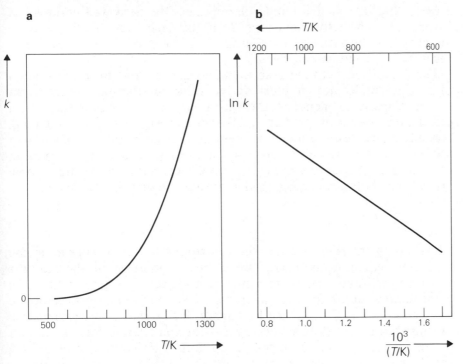

Figure 3.2 Variation of the rate constant with temperature. a. k plotted directly against T (in kelvins). b. $\ln k$ plotted against reciprocal temperature. T^{-1} has been multiplied by 10^3 to give a more convenient number scale. The reciprocal temperature scale is graduated directly in kelvins along the top margin. (This axis shows the recognized way of writing $10^3/T$ with T expressed in kelvins).

which the system is stable. The hurdle is illustrated for a hypothetical chemical reaction in Figure 3.3. It arises because the only pathway leading from reactants to products necessarily passes through a higher-energy **transition state**. For a reaction involving the rupture of one bond and the establishment of a new one, this involves the formation of

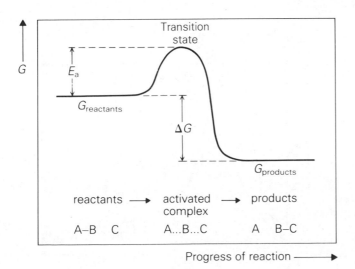

Figure 3.3 Energetics of a hypothetical reaction:

$$AB + C \rightarrow A + BC$$

The vertical axis represents the free energy of the system. E_a is the activation energy, which is released again (except in the case of endothermic reaction, when only a part is released) on completion of the molecular reaction.

a less stable intermediate molecular species (the **activated complex** in Figure 3.3). The activation energy E_a in the Arrhenius equation may be visualized as the 'height' of the hurdle (in free energy units) relative to the initial reactant assemblage.

For a collision between reactant molecules to lead to the formation of product molecules, it must derive sufficient thermal energy from the participants to generate the activated complex. From theoretical calculations one can predict the kinetic energy distribution among reactant molecules at a given temperature T: the results are illustrated for two temperatures in Figure 3.4. One can show that the proportion of molecular encounters involving kinetic energies greater than some critical threshold energy E_x (shaded areas in Figure 3.4) is given by

$$n \propto e^{E_x/RT} \tag{3.6}$$

The term on the right is called the **Boltzmann factor**. It appears in the Arrhenius equation as a measure of the proportion of the reactant molecule collisions that possess sufficient energy (i.e. greater than E_a) at the temperature T to reach the transition state and thereby to complete the reaction. The form of the Boltzmann factor shows that an increase of temperature will shift the energy distribution towards higher energies (Figure 3.4), so that a greater proportion of reactant molecules will collide with energies exceeding E_a and surmount the energy hurdle. In other words the reaction rate will increase. But reducing the temperature will inhibit the reaction, because $-E_a/RT$ becomes a larger negative number and makes the Boltzmann factor and therefore k smaller.

The activation energy indicates how sensitive the rate of a reaction will be to changes in temperature, and measurement of this temperature-dependence provides the means of determining E_a experimentally. (Because it is not a net energy change between reactants and products,

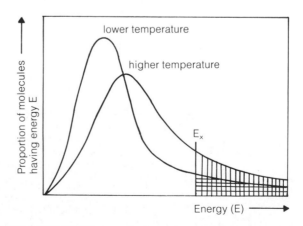

Figure 3.4 Molecular kinetic-energy distribution for two temperatures. The shaded areas show the portions of each distribution which lie above a specified energy E_x. This proportion is greater at the higher temperature.

Box 3.4

What activation energy means on the atomic scale

In a chemical reaction

$$AB + C \rightarrow A + BC$$

(Figure 3.3), the established A–B bond must be weakened (stretched) before a new bond (B–C) can begin to form. The energetics of the A–B bond are shown in diagram a. The process begins with AB in its most stable configuration. Energy is required to stretch the A–B bond to the stage when formation of the B–C bond becomes an equally probably outcome (i.e. formation of the activated complex A...B...C); this energy input constitutes the activation energy E_a (Figure 3.3). The whole reaction from AB to BC can be visualized by considering the energy–distance curves of both molecules, diagrams a and b back to back as in diagram c. Note that the AB bond need not be completely broken before the BC molecule can begin to form.

The explanation of the activation energy in reactions between ionic compounds is slightly different, but is still associated with the need to disrupt one arrangement of atoms or ions before another more stable arrangement can be adopted.

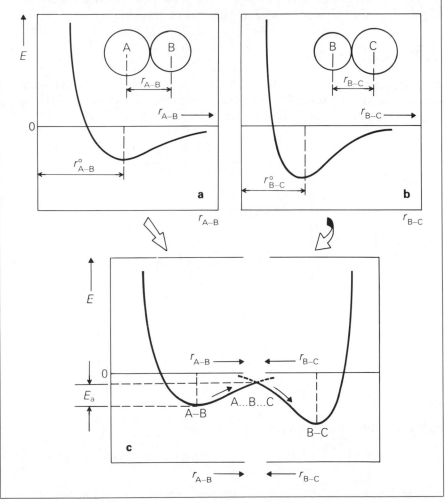

like ΔH, it cannot be measured calorimetrically.) Transforming both sides of Equation 3.5 into natural logarithms, one gets:

$$\ln k = \ln A - \frac{E_a}{R} \cdot \frac{1}{T} \tag{3.7}$$

which has the form

$$y = c + mx$$

This form of the Arrhenius equation shows a straight-line relationship between in k and $1/T$ (T must be expressed in kelvins), as shown in Figur 3.2b. The slope of this line is $-E_a/R$. The numerical value of the activation energy can therefore be established by repeating the rate experiment at a number of pre-determined temperatures, and plotting the rate constants obtained in the form shown in Figure 3.2b. Such a diagram is called an **Arrhenius plot**. A geological example is shown in Figure 3.5. See also exercise 2 at the end of the chapter.

3.3 Diffusion

Any process that requires the input of thermal energy to surmount an energy hurdle, a **thermally activated process**, will show an Arrhenius-type temperature-dependence (Equation 3.5 and Figure 3.2). This

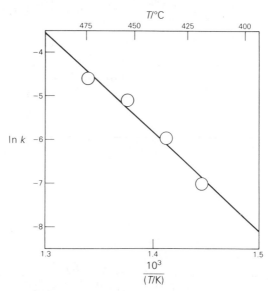

Figure 3.5 Arrhenius plot of the geochemical reaction:

$$NaAlSi_2O_6.H_2O + SiO_2 \quad NaAlSi_3O_8 + H_2O$$
analcime   quartz   albite   vapour

in NaCl solution (data from A. Matthews (1980) *Geochim. Cosmochim. Acta*, **44**, 387–402).

property is characteristic of a number of physical processes in geology in addition to geochemical reactions.

When a component is unevenly distributed in a phase – solid, liquid or gas – so that its concentration in one part is higher than in another, there is a net atom-by-atom migration, called **diffusion**, of the component 'down' the concentration gradient. This diffusion will lead eventually to a homogeneous distribution (labelled t_∞ in Figure 3.6a). In fact, atoms will diffuse even through a homogeneous substance, but only where there is a concentration gradient is a net flow of chemical components observed.

Imagine a crystal with an abrupt internal discontinuity in the concen-

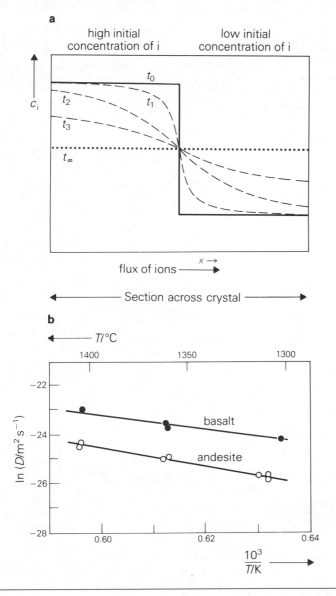

Figure 3.6 a. The diffusion of ions in response to a non-uniform concentration distribution. b. Arrhenius plot of diffusion coefficients for the diffusion of cobalt ions in silicate melts (after R. K. Lowry, P. Henderson and J. Nolan (1982) *Contr. Mineral. Petrol.*, **80**, 254–61). The activation energies measured from the slopes of the two graphs are:

basalt: 220 kJ mol^{-1}
andesite: 280 kJ mol^{-1}

The units of D are m^2 s^{-1}.

Box 3.5

How concentration changes with time – Fick's Second Law

Equation 3.8 allows us to calculate how much of component i will be transported through a window of unit area during a specified interval of time, if we know the concentration gradient. As a result of this flux, the spatial distribution of the chemical species i will change. Suppose we ask a slightly different question: how rapidly does the *concentration* of i at point x change with time? This can be calculated from **Fick's Second Law**. If we denote the concentration gradient dc_i/dx (in [mol m^{-3}] m^{-1}) by the symbol q_i, the Second Law may be written:

$$\frac{dc_i}{dt} = D_i \frac{dq_i}{dx}$$

This equation says that the concentration will change quickly if the concentration gradient itself varies rapidly from one point to another (e.g. the points labelled B in diagram b

below). On the other hand, if the concentration gradient is uniform, as at point A, then the concentration will not change with time.

The figure illustrates three possible scenarios; the first one, in diagram a, has a uniform gradient and the other two, in diagram b, show variable gradients (equivalent to curves t_1 and t_3 in Figure 3.6a).

We must remember that q_i is itself a derivative. A mathematician would write Fick's Second Law in the form:

$$\frac{dc_i}{dt} = D_i \frac{d^2c_i}{dx^2}$$

where d^2c_i/dx^2 is the notation for a *second derivative* (the derivative of a derivative – see Appendix B and Waltham, 1994), a measure of the *curvature* of the concentration profile.

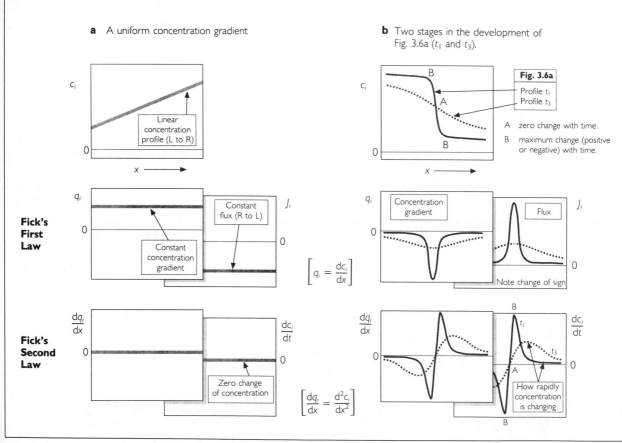

a A uniform concentration gradient

b Two stages in the development of Fig. 3.6a (t_1 and t_3).

Fig. 3.6a
Profile t_1
Profile t_3

A zero change with time.
B maximum change (positive or negative) with time.

Fick's First Law

$$\left[q_i = \frac{dc_i}{dx} \right]$$

Note change of sign

Fick's Second Law

$$\left[\frac{dq_i}{dx} = \frac{d^2c_i}{dx^2} \right]$$

tration of a component i, as shown in Figure 3.6a. Suppose the crystal is maintained at a constant temperature T high enough for solid-state diffusion to occur at a significant rate. If we were to measure the distribution of component i on several successive occasions t_1, t_2, t_3, we would see the development of a progressively smoother concentration profile, leading eventually to a uniform distribution of i (at t_∞ in Figure 3.6a). These changes point to a net flux of component i from left to right through the plane of the original discontinuity. The magnitude of the flux in moles per second will depend on the surface area of the interface. One therefore expresses the flux $\mathcal{J}_i$ as the amount of component i (in moles) that migrates through a unit area of the plane per second, so the units are 'moles per square metre per second' ($\mathrm{mol\,m^{-2}\,s^{-1}}$). Common sense suggests that the flux will depend on the steepness of the concentration gradient (i.e. no net flux will occur if the concentration is uniform and the gradient zero). Theory shows that the relationship is very simple:

$$\mathcal{J}_i = -D_i \frac{\mathrm{d}c_i}{\mathrm{d}x} \qquad (3.8)$$

This equation is called **Fick's First Law of Diffusion**. The negative sign indicates that the net flux flows down the concentration gradient (i.e. towards the right in Figure 3.6a). The constant D_i is called the **diffusion coefficient** for the species i in the crystal concerned (at a given temperature T). Like the rate constant for a chemical reaction, it has a large value for rapid diffusion and a small value for slow diffusion. If the units of concentration c_i are 'moles per cubic metre' ($\mathrm{mol\,m^{-3}}$), the concentration gradient $\mathrm{d}c_i/\mathrm{d}x$ will be expressed in units of moles per cubic metre, per metre in the x direction ($[\mathrm{mol\,m^{-3}}]\,\mathrm{m^{-1}} = \mathrm{mol\,m^{-4}}$) from which point it is simple to show that the units of D_i must be $\mathrm{m^2\,s^{-1}}$.

Many experimental determinations of diffusion coefficients have been made for various elements in a range of silicate materials. Figure 3.6b shows how the diffusion coefficient for cobalt (D_{Co}) changes with temperature and composition for silicate melts of two compositions analogous to natural basalt and andesite. The data are presented in the from of an Arrhenius plot, in which they lie on straight lines. Thus one can express D in terms of an equation similar to Equation 3.6:

$$D = D_0 e^{-E_a/RT} \qquad (3.9)$$

or

$$\ln D = -\frac{E_a}{R} \cdot \frac{1}{T} + \ln D_0 \qquad (3.10)$$

These equations are identical to Equations 3.5 and 3.7, except that the pre-exponential factor is written D_0. Thus although we think of diffusion as a physical phenomenon, it resembles a chemical reaction in being governed by an activation energy. Like a person caught up in a dense crowd, the diffusing atom must jostle and squeeze its way through the voids of the melt (Chapter 8); pushing through from one structural site to the next presents an energy hurdle which only the more energetic atoms can surmount.

Diffusion coefficients differ from one element to another (Co as opposed to Cr, for example) for diffusion in the same material at the same temperature. Figure 3.6b shows that the diffusion coefficient for one element in a melt will also vary with the composition of the melt (Box 8.2).

3.3.1 Solid-state diffusion

From the point of view of diffusion, a crystalline solid differs from a melt in several important respects. Atoms diffusing through a silicate melt encounter a continuous, isotropic medium (D is independent of direction) that is a relatively disordered structure. Most crystalline solids on the other hand are polycrystalline aggregates that offer two routes for diffusion, within and between crystals.

VOLUME (INTRA-CRYSTALLINE) DIFFUSION

Volume diffusion through the three-dimensional volume of the constituent crystals is similar in general terms to diffusion through a melt. The main difference is that the crystals have more closely packed, ordered atomic structures than melts (Box 8.2), and diffusion through them is much slower (lower D). Figure 3.7 contrasts the diffusion behaviour of similar metals in melts and in olivine crystals (in which the diffusion coefficients are less by a factor of about 10^5). Notice that diffusion in olivine takes place more readily along the crystallographic z-axis than along the y-axis. Crystallographic direction is an important factor in diffusion through anisotropic crystals.

GRAIN-BOUNDARY (INTER-CRYSTALLINE) DIFFUSION

Grain-boundary diffusion exploits the structural discontinuity between neighbouring crystals as a channel for diffusion. This is much more difficult to quantify because the rate of diffusion depends on:

(a) the grain size of the rock: in a fine-grained rock, the total area of grain boundaries is larger in relation to the total volume of the rock, and diffusion will be easier;

(b) the microscopic characteristics of the grain boundaries, e.g. the presence or absence of water.

Grain-boundary diffusion appears to be much more rapid in most

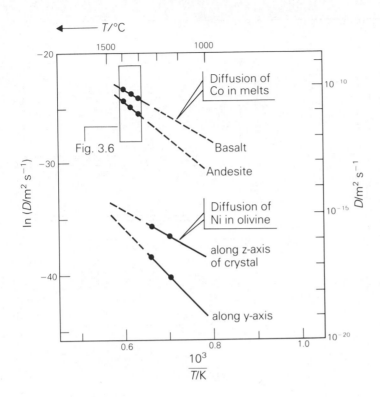

Figure 3.7 Comparison of volume diffusion rates in silicate melts and olivine crystals. Data from Henderson (1982).

circumstances than volume diffusion. Grain boundaries provide the main conduit through which volatile species penetrate the volume of a rock during metamorphism and hydrothermal alteration. Open fractures and faults do not occur in the deep crust owing to the high temperature and confining pressure. Under these conditions fluid migration is concentrated along **shear zones**, where deformation has led to a local reduction of grain size and often to a strong foliation, both of which promote fluid movement by an increase in grain-boundary area.

Box 3.6 illustrates how diffusion studies can be applied to estimate the cooling rates experienced by iron meteorites.

3.4 Viscosity

The flow of silicate melts is another physical process that is very sensitive to temperature. A liquid flows in response to shear stress exerted upon it, which generates a velocity gradient in the liquid. The velocity of water flowing in a pipe of radius r, for example, increases from zero at the wall to a maximum v at the centre, an average gradient of v/r. For most liquids the velocity gradient is proportional to the applied shear stress σ:

Box 3.6

Diffusion and cooling rate; applications to meteorites

At temperatures below 900 °C, the solid solution between iron metal and nickel metal is divided by a two-phase region below a solvus (diagram a). Many iron meteorites exhibit complex exsolution intergrowths of two separate Fe–Ni alloys, kamacite and taenite, developed during cooling through this solvus (Box 10.3).

Consider the cooling of alloy X which at 900 °C is a homogeneous metallic solid solution. At about 700 °C the alloy encounters the solvus, below which taenite becomes supersaturated with Fe, and expels it in the form of platelets of a separate kamacite phase. The upper inset shows a profile of Ni content in a representative cross-section (on the scale of a few mm) at 680 °C: slender lamellae of low-Ni kamacite have appeared in the initially homogeneous taenite.

The Fe–Ni solvus is unusual in that both limbs slope in the same direction (cf. Figure 2.6). As the temperature falls, therefore, both immiscible phases grow more Ni-rich. The lever rule (Box 2.3) indicates that their relative proportions must also change: the two lower insets show that kamacite spreads at the expense of taenite, forming ever thicker plates.

This process depends on the diffusion rates of Fe and Ni atoms in the two phases. Measurements show that diffusion is slower in taenite than in kamacite, and Ni is expelled from kamacite more rapidly than it can diffuse into the interior of the adjacent taenite crystal. Consequently the Ni distribution in taenite lamellae develops an M-shaped profile, with Ni concentrated at the edges. The more rapidly the meteorite is cooled, the more pronounced is the dip in the Ni profile. Calculations based on diffusion profiles (diagram b) and Fick's Second Law (Box 3.5) allow estimation, from the shape of the M-profiles in iron meteorites, of the cooling rates they experienced during the early development of the solar system. These estimates, commonly between 1 °C and 10 °C per million years, suggest that iron meteorites are derived from relatively small parent bodies (diameter < 400 km); large planetary bodies would cool more slowly (Hutchison, 1983, in references to Chapter 10).

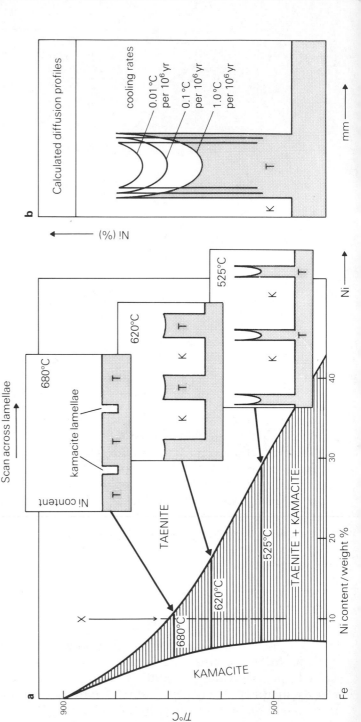

$$\text{velocity gradient} = \frac{\sigma}{\eta} \qquad (3.11)$$

The constant η is called the **viscosity** of the liquid. Note that viscosity is *inversely* related to a liquid's capacity to flow. A low viscosity indicates a runny liquid, a high value a 'stiff' one. Methods by which the viscosity of silicate melts may be measured are summarized by Holloway and Wood (1988). To put magnitude in perspective, the values plotted for basalt vary from 7.5 to $150\,\mathrm{N\,s\,m^{-2}}$; for comparison the viscosity of water at $25\,^\circ\mathrm{C}$ is $0.001\,\mathrm{N\,s\,m^{-2}}$.

Figure 3.8 shows that the viscosity of silicate melts, like that of motor oils, is strongly temperature-dependent.

$$\frac{1}{\eta} = \frac{1}{\eta_0}e^{-E_a/RT} \qquad (3.12)$$

The flow of a liquid, like diffusion, requires atoms or molecules to jostle past each other, surmounting energy hurdles as they do so. The activation energies for the flow of silicate melts are generally higher than for diffusion.

The deformation of crystalline rocks, though a much more complicated process, shows a similar dependence on temperature:

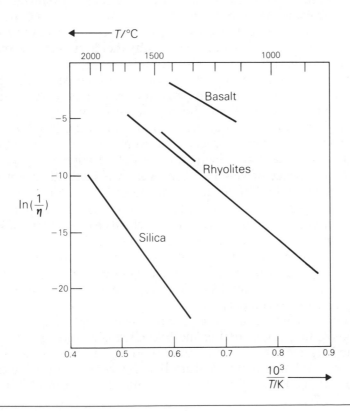

Figure 3.8 Arrhenius plot of viscosities (in $\mathrm{N\,s\,m^{-2}}$) for some silicate melts. (Data from C. M. Scarfe (1977) *Canad. Mineral.*, 15.) Viscosity is the inverse of the flow rate-constant so it has been plotted here in reciprocal form. The slope of the graphs increases with silica content, reflecting an increase in activation energy as follows:

basalt: $230\ \mathrm{kJ\ mol^{-1}}$
rhyolite: $350\ \mathrm{kJ\ mol^{-1}}$
silica: $500\ \mathrm{kJ\ mol^{-1}}$

$$\text{strain rate} = (Ae^{-E_a/RT})\sigma^N \qquad (3.13)$$

where σ is again shear stress, and N is a constant depending on details of the deformation process (its value is typically about 8). The factor in brackets is analogous to inverse viscosity. It is not given that name because σ, instead of appearing in linear form, is raised to the power N in Equation 3.13. Activation energies are, however, in the same range as those for viscous flow of melts (see the caption for Figure 3.8).

3.5   Persistence of metastable minerals; closure temperature

Paradoxically, chemical equilibrium in geological processes would be of less interest to the petrologist were it not for the intervention of disequilibrium. Consider an argillaceous rock undergoing metamorphic recrystallization in conditions corresponding to the kyanite–sillimanite phase boundary (Figure 2.1). Once equilibrium is achieved, kyanite and sillimanite will coexist stably in the rock. If, following uplift and erosion, the rock is found at the surface still containing kyanite and sillimanite, their coexistence and textural relationship will testify to the high-temperature conditions under which the rock crystallized and will give the petrologist an indication of the conditions of metamorphism. Yet the current state of the rock is plainly one of disequilibrium, as sillimanite is now well outside its stability field; it remains only as a metastable relic of a former state of equilibrium.

Metamorphic equilibrium seems generally to keep pace with changing conditions more effectively in prograde reactions (those involving a temperature increase) than during the waning stages of regional metamorphism. One reason is that volatile constituents such as water, whose presence accelerates recrystallization, will be more abundant during the prograde stage owing to the dehydration reactions taking place. In the absence of volatiles, however, it is likely that temperature asserts the dominating influence on metamorphic reaction rates.

As metamorphic or igneous rocks cool, they pass into a temperature range of progressive kinetic paralysis, in which reaction rates fall behind changing circumstances (as illustrated in Box 3.6). Eventually a temperature is reached (dependent on the reaction being considered) at which diffusion rates become substantially slower that the cooling rate of the rock, and reaction has effectively ceased. This is called the **blocking** or **closure temperature**.

The closure temperature is an important concept in geochronology. In the potassium–argon dating method (Box 9.2) the age of a rock is established by measuring the minute amount of ^{40}Ar (argon is a gas) that has accumulated in a potassium-bearing mineral form the decay of ^{40}K. The method therefore depends critically on the ability of the relevant mineral grains to retain this intra-crystalline argon component. At temperatures above the closure temperature for argon diffusion, the ^{40}Ar atoms diffuse to grain boundaries and escape. A K–Ar age-determination on an igneous rock therefore records not the age of intrusion or eruption, but the date when the rock had cooled to temperatures low enough for the rate of diffusion of ^{40}Ar out of the grains of the potassium mineral to be insignificant. A later phase of metamorphism in which the rock is heated above the closure temperature would discharge the accumulated ^{40}Ar, and the isotopic clock would thereafter record a 'metamorphic age', indicating the time when the rock body had again cooled below the closure temperature.

3.6 Review

A chemical reaction can be visualized in free energy terms as a journey leading from one valley, the domain of the reactants, to another where the product species form the dominant population (Figure 3.3). The route from one valley to the other leads across a high pass in free energy space, the 'transition state', and reactant molecules that cannot summon sufficient energy to traverse this high ground will not reach their destination. The key factor is the availability of energy: if energy is plentiful, the traffic over the pass will be heavy (the reaction rate will be high); if energy is hard to come by, many reactant molecules will be forced to turn back before attaining the pass.

What sources of energy drive geochemical reactions? For reactions taking place in the Earth's interior, reactants must rely on the kinetic energy possessed by their constituent atoms and molecules. Accordingly the temperature plays a crucial role in determining the rates of geochemical reactions, as enshrined in the Arrhenius equation. Reactions involving silicate minerals are characterized by large activation energies, and therefore only at high temperatures is there a significant population of reactant 'molecules' possessing sufficient kinetic energy to surmount the activation energy hurdle (Figure 3.4). Disequilibrium textures in many rocks (Box 3.1) testify to the rapid slowing-down of chemical reactions as temperature falls, and at surface temperatures disequilibrium is the rule rather than the exception: the persistence of Fe^2 minerals on the Earth's surface, where atmospheric oxygen makes Fe^3 the stable form of iron, is one obvious example. Both the flow of silicate melts (Figure 3.8) and diffusion are also strongly temperature-dependent (Figures 3.6 and 3.7), suggesting they too involve an activation step, The marked slowing-down of diffusion with falling temperature is an essential requirement for radiometric dating.

Many reactions in the atmosphere, however, rely on another energy source, the Sun. Solar photons, particularly those of UV wavelengths, are sufficiently energetic to tear apart chemical bonds in molecules like O_2, O_3, NO_2, H_2O_2, $CFCl_3$ and $HCHO$ (formaldehyde). The result of such **photodissociation** reactions is often the formation of **free radicals** such as $O\cdot$, $HO\cdot$, $H\cdot$, $HO_2\cdot$, $NO_3\cdot$ $Cl\cdot$, $ClO\cdot$ and $HCO\cdot$. Free radicals possess unpaired electrons (represented by the symbol '$\cdot$') and are highly reactive (Chapter 7): through photodissociation they have already reached the energy 'pass' and thus they are capable of initiating gas reactions in a manner unrestrained by considerations of activation energy. For example:

$$CH_4 + HO\cdot \rightarrow H_3C\cdot + H_2O \tag{3.14}$$

This reaction illustrates, incidentally, the vital role of the hydroxyl radical $HO\cdot$ as an atmospheric cleansing agent, removing many forms of pollution from the air that we breathe.

3.7   Bibliography

Atkins, P. W. (1994) *Physical chemistry*, 5th edn, Oxford University Press, Oxford.

Henderson, P. (1982) *Inorganic geochemistry*, Pergamon, Oxford.

Holloway, J. R. and Wood, B. J. (1988) *Simulating the Earth – experimental geochemistry*, Unwin Hyman, Boston.

Lasaga, A. C. and Kirkpatrick, R. J. (1981) *Kinetics of geochemical processes*, Mineralogical Society of America (Reviews in Mineralogy, volume 8).

MacKenzie, W. S., Donaldson C. H. and Guilford, C. (1982) *Atlas of igneous rocks and their textures*, Longman, London.

Waltham, D. (1994) *Mathematics: a simple tool for geologists*, Chapman & Hall, London.

Yardley, B. W. D. (1989) *An introduction to metamorphic petrology*, Longman, Harlow.

3.8   Exercises

1   The table below gives the Geiger counter count-rate at various times during an experiment on radioactive iodine-131 (^{131}I). Treating the decay of ^{131}I as a chemical reaction in which the count rate may be taken as being a measure of the concentration of ^{131}I, confirm that the reaction is first-order and determine the decay constant. From the linear graph you have drawn, calculate the half-life of ^{131}I (Box 3.2).

   ^{131}I is a volatile **fission product** (Box 10.2), formed in fuel rods during the operation of nuclear reactors, and represents a serious radiation hazard in the event of a gas leak from a reactor containment vessel. Calculate the time it would take for the radioactivity of ^{131}I to decay to one-hundredth of its value on escaping into the atmosphere.

Time (hours)	Counts (sec^{-1})
0	18 032
25	16 410
50	15 061
100	12 590
200	8 789
300	6 144
400	4 281
500	3 002

2 There is a rule of thumb which says that the rate of many room-temperature chemical reactions approximately doubles when the temperature is increased by 10 °C. Calculate the activation energy for a reaction that exactly conforms to this relationship. (Gas constant $R = 8.3143 \text{J K}^{-1} \text{mol}^{-1}$.)

3 Use the viscosity measurements given below to verify that the Arrhenius equation is applicable to the flow of a silicate melt, and determine the activation energy. (Hint: viscosity is the resistance to flow, to which flow rate is inversely proportional.)

Temperature (°C)	Viscosity of rhyolite melt (N s m^{-2})
1325	2042
1345	1585
1374	1097
1405	741

(Gas constant as given in Exercise 2.)

4 Calculate the half-life of ^{87}Rb ($\lambda_{^{87}\text{Rb}} = 1.42 \times 10^{-11}$ year^{-1}). What percentage of the ^{87}Rb incorporated in the Earth 4.6×10^9 years ago has decayed to ^{87}Sr?

4

Aqueous solutions in geology

The great importance of water and aqueous solutions on the surface of the Earth barely needs pointing out. Water is the principal agent of erosion and of the transportation of eroded materials, either by mechanical or chemical means. The world's oceans are the primary medium for sedimentation, they act as a global chemical repository for many substances of geological significance, and they play a crucial part in moderating the climate and supporting life on the planet. Water also has important functions in the Earth's interior: ore transport, rock alteration and metamorphism all involve the migration of hot aqueous fluids through the crust. The chemistry of aqueous solutions, the subject of this chapter, is a vital element of all these geological processes.

The basic jargon of solution chemistry is reviewed in Appendix C and in the Glossary. Some of the important properties of water as a solvent are reviewed in Box 4.1.

4.1 Ways of expressing concentration

SOLUTIONS

The composition of a solution can be expressed by stating the **concentration** of each of the solute species present. There are several ways of doing this. The obvious one is to state the mass of each solute per unit volume of solution, in units such as $g\,dm^{-3}$. But in considering the properties of the solution, it is the number of **moles** of each substance present that matters, rather than the weight of solute in solution. Dividing the concentration C of each compound in the solution (expressed in $g\,dm^{-3}$) by its relative molecular mass M_r gives the **molarity**, c:

$$c = C/M_r\,mol\,dm^{-3}$$

Box 4.1

The special properties of water

Water is so commonplace that one tends to overlook to how unusual its properties are in comparison with other liquids. The water molecule is bent (Chapter 7), with both hydrogen atoms occurring on the same side. As oxygen attracts electrons more strongly than hydrogen, a slight excess of electronic (negative) charge $\delta-$ exists at the oxygen 'end' of the molecule, leading to a corresponding deficiency (a net positive charge $\delta+$) on the hydrogen atoms at the other end (Figure 7.9). This polarity is responsible for most of the properties peculiar to water:

(a) Molecules of water in the liquid and solid states are loosely bonded to each other by an electrostatic attraction between the negative end of one molecule and the positive end of a neighbouring one. This **hydrogen bonding** (Chapter 7) lies behind many of water's unique properties.

(b) Hydrogen bonding gives liquid water a very high specific heat, a high latent heat of vaporization (because water molecules are difficult to separate and disperse into a gas phase), and an unusually large liquidstate temperature range (100 °C). These thermal properties of water give it a heat-exchange capacity of great climatic significance on the Earth, as the moderating effect of the oceans on seaboard climate illustrates. It

takes only 2.5 m depth of ocean water to match the heat capacity of the entire atmospheric column above.

(c) Hydrogen bonding in the liquid state is also responsible for making liquid water denser than ice at temperatures close to 0 °C (for the consequences of which see Box 2.2). Water therefore expands on freezing, leading to the important erosional processes of frost-shattering and heaving. Water also has an unusually high surface tension, resulting in strong capillary penetration of pore waters.

(d) The polar nature of the water molecule makes water an attractive environment for ions, and gives water its unique capacity as a solvent of ionic compounds. Water molecules attracted by the electrostatic field of each ion cluster in a loose layer around it, aligning their polarity according to the ion's charge. This association of water molecules around a dissolved ion (an **ion-dipole interaction** – Chapter 7) is called **hydration**. The combined electrostatic attraction of the polar water molecules lowers the ion's potential energy and stabilizes it in solution. (Many ionic compounds remain hydrated when they crystallize from solution, forming hydrated salts such as the mineral gypsum, $CaSO_4 \cdot 2H_2O$.) Water is thus the prime example of a **polar solvent**.

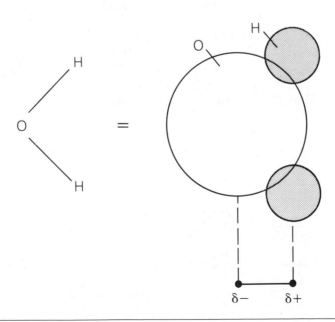

$\delta-$ $\delta+$

Aqueous geochemists measure solute concentrations using a slightly different quantity called the **molality**, m. The molality m_i of species i in an aqueous solution is the number of moles of i per kilogram of water (solvent, not solution). There are sound practical and thermodynamic reasons, however, for expressing molality in **dimensionless** form (a form in which the units of measurement cancel out). For this purpose we use the **activity** a_i, which for **dilute solutions** is numerically equal to the molality:

$$a_i = m_i/m^\circ \qquad (4.1)$$

where m° is the **standard molality**, equal to $1\,\mathrm{mol\,kg^{-1}}$. This artifice removes the necessity of associating units with complex formulae like Equation 4.8.

SOLIDS

In geology, one often deals with heterogeneous reactions between aqueous solutions and solid or gaseous phases. The concentration of a component in a solid phase is best expressed in terms of its **mole fraction** (X) in the solid.

Consider a sample of olivine that has been found to contain 65.0% by mass of Mg_2SiO_4 (the forsterite end-member, p. 22) and 35.0% of Fe_2SiO_4 (the fayalite end-member). To calculate the mole fractions of these components (assuming no other components are present), one must first establish their relative molecular masses M_r from the following relative atomic masses:

$$Mg = 24.31 \quad Si = 28.09 \quad Fe = 55.85 \quad O = 16.00$$

The relative molecular mass (RMM) of forsterite is therefore $(2 \times 24.31) + 28.09 + (4 \times 16.00) = 140.71$, and therefore 1 mole of Mg_2SiO_4 will weigh $140.71\,\mathrm{g}$. The RMM of fayalite is $(2 \times 55.85) + 28.09 + (4 \times 16.00) = 203.79$.
Thus 100 g of the olivine will contain:

(a) 65.0 g of Mg_2SiO_4 = 65.0 / 140.71 = 0.4619 moles Mg_2SiO_4;
(b) 35.0 g of Fe_2SiO_4 = 35.0 / 203.79 = 0.1717 moles Fe_2SiO_4.

The mole fraction of Mg_2SiO_4 in this olivine is therefore:

$$X_{Mg_2SO_4} = \frac{\text{moles of } Mg_2SiO_4}{\text{total moles}} = \frac{0.4619}{0.4619 + 0.1717} = 0.7290 \qquad (4.2a)$$

Likewise the mole fraction of Fe_2SiO_4 in the olivine is:

$$X_{Fe_2SiO_4} = \frac{\text{moles of } Fe_2SiO_4}{\text{total moles}} = \frac{0.1717}{0.4619 + 0.1717} = 0.2710 \qquad (4.2b)$$

In this case, checking whether the mole fractions add up to 1.0000 provides a precaution against arithmetic error.

GASES

The concentration of a component in a gas phase is expressed in terms of its **partial pressure**. For an ideal gas, the partial pressure of component A is $P_A = X_A P$ where X_A is the mole fraction of A in the gas and P is the total pressure. In some circumstances the fugacity f_A is used in place of partial pressure (p. 218).

4.2 Equilibrium constant

In this chapter we are concerned with reactions in solution that can proceed in either direction, depending on the circumstances. When such reactions occur, the forward reaction

$$A + B \rightarrow C + D \tag{4.3}$$

has to compete with the reverse reaction

$$A + B \leftarrow C + D \tag{4.4}$$

and together they arrive at a state of **equilibrium** in which the rate of one is exactly matched by that of the other. This is written:

$$A + B \rightleftharpoons C + D \tag{4.5}$$

Once equilibrium is achieved, there is no further change in any of the concentrations. The composition of this steady state can be summarized by the **equilibrium constant** K:

$$K = \frac{a_C a_D}{a_A a_B} \tag{4.6}$$

The value of K is constant for a particular reaction at a specified temperature. It indicates at which point on the path from '100% reactants' (A + B) to '100% products' (C + D) the reaction settles into equilibrium. The system, if disturbed, will always readjust to the same position of equilibrium. For example, if more A is added to the equilibrated solution represented by Equation 4.5 (i.e. if a_A is increased), A and B will react together to a greater extent, thereby reducing a_B (and the new a_A) and increasing a_C and a_D. These individual activity adjustments occur in such a way that the ratio $a_C a_D / a_A a_B$ resumes the original value, K – the value it had before the disturbance occurred. Altering the temperature, however, will cause the value of K to change (Box 4.2).

Box 4.2

The effect of temperature on equilibrium constants

Like the equilibria considered in Chapter 2, the positions of aqueous equilibria vary with temperature (and to a much smaller extent with pressure). An equilibrium constant is therefore only constant for a particular set of physical conditions. The temperature variation is important, considering the wide range of temperatures at which aqueous solutions are encountered in the geological world (from $0\,°C$ to hundreds of degrees).

Raising the temperature of a system in equilibrium will alter the equilibrium constant in a direction which favours the higher enthalpy side of the equilibrium (Le Chatelier's principle – Chapter 2). Consider for example the solubility of a salt like $BaSO_4$:

$$BaSO_4 \rightleftharpoons Ba^{2+} + SO_4^{2-}$$
$$\text{solid} \qquad \text{saturated solution}$$

This reaction has a positive ΔH (the reaction is endothermic). As enthalpy is higher on the right-hand side, raising the temperature will increase the equilibrium constant (Equation 4.13), which accords with the general experience that salts are more soluble in hot water than cold.

During the 1980s a programme of exploration in the Pacific Ocean using the manned submersible *Alvin* located active, chimney-shaped hydrothermal vents, which discharge jets of water at temperatures up to $350\,°C$. Such vents have since been identified on mid-ocean ridges in all of the major oceans. The hot, mildly acid fluids discharged from them are saturated with metal sulphides and other salts, which are immediately precipitated on coming into contact with cold neutral sea water. The fine sulphide precipitates form dense black clouds in the water, owing to which these vents have become known as 'smokers'. (Edmonds and von Damm, 1983; Richards and Strens, 1985.)

Gases on the other hand become less soluble at higher temperatures. The expulsion of dissolved CO_2 as water is heated causes the deposition of the familiar 'fur' inside kettles and heating pipes, particularly in hard-water areas. Such waters contain calcium bicarbonate which, with the expulsion of dissolved CO_2 as the water is heated, forms the less soluble calcium carbonate:

$$Ca^{2+} + 2HCO_3^- \rightarrow CaCO_3 + H_2O + CO_2$$
$$\text{hard water} \qquad \text{solid} \quad \text{water} \quad \text{expelled}$$
$$\text{(cold)} \qquad \text{'fur'} \quad \text{(hot)} \quad \text{gas}$$

Some solids also have inverse solubility/temperature relationships, e.g. the anhydrite ($CaSO_4$) from which smoker chimneys are initially built. The anhydrite is deposited from sea water as it is heated by the hot vent.

The algebraic form of the equilibrium constant reflects the nature of the reaction equation. If we were to look at a more complicated equilibrium, for example one in which b molecules of species B react with c molecules of species C like this:

$$bB + cC \rightleftharpoons dD + eE + fF \tag{4.7}$$

we would find that the equilibrium constant had the form:

$$K = \frac{a_D{}^d \cdot a_E{}^e \cdot a_F{}^f}{a_B{}^b \cdot a_C{}^c} \tag{4.8}$$

4.2.1 Solubility and the solubility product

To see how an equilibrium constant works, consider the solubility of various substances in water. The salt calcium fluoride, CaF_2 (which, in crystalline form, is the familiar vein mineral fluorite), is only very slightly soluble in cold water. $0.017\,g$ of CaF_2 will dissolve in $1\,kg$ of water at $25\,°C$, corresponding to a molality of $0.000\,22\,mol\,kg^{-1}$. (The relative molecular mass of CaF_2 is $40.08 + (2 \times 19.00) = 78.08$.) This quantity is the **solubility** of CaF_2 at this temperature. Adding further solid CaF_2 to the system will cause no increase in the concentration in solution, no matter how long we wait. Such a solution, which has reached equilibrium with solid CaF_2 left over, is said to be **saturated** with this component.

Calcium fluoride, being an ionic compound (Chapter 7), is completely dissociated in aqueous solution into calcium and fluoride **ions** (Ca^{2+} and F^- respectively). The calcium atom loses two electrons and becomes a doubly charged positive ion or **cation**. Each fluorine atom (of which there are twice as many as calcium) captures one extra electron to become a singly charged negative ion or **anion**. This process can be written as a chemical reaction leading to an equilibrium:

$$CaF_2 \rightleftharpoons Ca^{2+} + 2F^- \tag{4.9}$$

crystal   *solution*

(Because that is a **heterogeneous equilibrium** like those considered in Chapter 2, one needs to specify in which phase each reactant or product resides when writing the reaction.) As dissolution proceeds, the concentration of Ca^{2+} and F^- ions in solution builds up, and increasingly the ions will react with each other to produce solid CaF_2 again (the reverse reaction). Saturation of the solution is an equilibrium in which the gross rate of dissolution of solid CaF_2 is equalled by the rate of precipitation from solution, so that no net change is observed. One can formulate an equilibrium constant for Equation 4.9:

$$K_{CaF_2} = \frac{a_{Ca^{2+}} \cdot (a_{F^-})^2}{X_{CaF_2}} \tag{4.10}$$

(Compare Equation 4.8.) Because the solid phase is pure CaF_2, the mole fraction $X_{CaF_2} = 1.00$. Thus:

$$K_{CaF_2} = a_{Ca^{2+}} \cdot (a_{F^-})^2 \tag{4.11}$$

where $a_{Ca^{2+}}$ and a_{F^-} are the activities of Ca^{2+} and F^- ions in a solution saturated with CaF_2.

This kind of equilibrium constant provides an alternative way of expressing the solubility of a slightly soluble salt (in this case CaF_2) in water, it is called the **solubility product** of CaF_2. It conveys the same information as the solubility, but in a different – and more versatile – form.

The value of K_{CaF_2} can be calculated from the solubility data given above. Reaction (4.9) tells us that one mole of solid CaF_2 dissolves (in sufficient water) to yield one mole of Ca^{2+} ions and two moles of F^- ions. Thus if 0.000 22 moles of CaF_2 can saturate 1 kg of water at 25 °C, the activities of Ca^{2+} and F^- in the saturated solution will be 0.000 22 and 0.000 44 respectively.

Therefore:

$$
\begin{aligned}
K_{CaF_2} &= a_{Ca^{2+}} \cdot (a_{F^-})^2 \\
&= 0.000\,22 \times (0.000\,44)^2 \\
&= 4.26 \times 10^{-11}
\end{aligned}
$$

It is often more convenient to write K_{CaF_2} in logarithmic form:

$$\log_{10} K_{CaF_2} = -10.4$$

or
$$K_{CaF_2} = 10^{-10.4}$$

or, by analogy with pH (Appendix C),

$$pK_{CaF_2} = 10.4$$

Solubility products are available in published tables (see references at the end of this chapter). Like all equilibrium constants they vary with temperature (Box 4.2).

If in a solution of CaF_2 the observed ion activity product

$$a_{Ca^{2+}} \cdot (a_{F^-})^2$$

has a value (e.g. 10^{-12}) that is numerically smaller than the solubility product K_{CaF_2} for the appropriate temperature, the solution is not saturated with CaF_2. Any solid CaF_2 introduced will be unstable, and will tend to dissolve. But if circumstances produce an ion activity product greater than $10^{-10.4}$ (say 10^{-8}) at 25 °C, the solution has become **supersaturated** with CaF_2, and will precipitate solid CaF_2 until the activity product has fallen to the equilibrium value indicated by the solubility product.

Table 4.1 shows the solubility products for a few important minerals. More data relevant to geochemistry are given by Krauskopf (1995).

Table 4.1 Solubility products (tabulated as pK (= $-\log K$) for 25 °C).

Halides			Carbonates*		
$PbCl_2$	4.8	(i.e. $K = 10^{-4.8}$)	$CaCO_3$	8.3	(calcite)
BaF_2	5.8		$BaCO_3$	8.3	
CuCl	6.7		$FeCO_3$	10.7	
AgCl	9.7		$MgCO_3$	6.5	
CaF_2	10.4				
			Sulphides		
			PbS	27.5	
Sulphates			HgS	53.3	
$BaSO_4$	10.0		ZnS	24.7	
$CaSO_4$	4.5				
$PbSO_4$	7.8				
$SrSO_4$	6.5		Phosphate		
			$Ca_5(PO_4)_3F$	60.4	(fluorapatite)

* Solubility dependent on pH and H_2CO_3 concentration.

INTERACTION BETWEEN IONIC SOLUTES: THE COMMON-ION EFFECT

We have been concerned up to now with solutions containing only a single salt (CaF_2). Natural waters, however, are mixed solutions containing many salts, and in such circumstances the question of solubility ceases to be a simple matter, because the solubility of CaF_2, for example, is now affected by contributions of Ca^{2+} and F^- from other salts present which contain these ions (such as $CaSO_4$ and NaF respectively).

Consider the salt barium sulphate, $BaSO_4$, which forms the mineral barite (another vein mineral). On dissolving in water, $BaSO_4$ ionizes as follows:

$$\underset{crystal}{BaSO_4} = \underset{solution}{Ba^{2+} + SO_4^{2-}} \qquad (4.12)$$

The solubility product at 25 °C is

$$K_{BaSO_4} = a_{Ba^{2+}} \cdot a_{SO_4^{2-}} = 10^{-10.0} \quad \text{(from Table 4.1).} \qquad (4.13)$$

The corresponding calcium salt $CaSO_4$ is more soluble at 25 °C:

$$K_{CaSO_4} = 10^{-4.5}$$

Suppose now we mix equal volumes of

(a) a saturated $BaSO_4$ solution ($BaSO_4$ activity 10^{-5}), and
(b) a $CaSO_4$ solution with an activity of $0.001 = 10^{-3}$ (which the reader can easily confirm is undersaturated with $CaSO_4$).

Mixing these solutions dilutes both $BaSO_4$ and $CaSO_4$ by a factor of two (the same amount of each salt now dissolved in twice the volume of water). We might therefore expect the mixed solution to be less than saturated with $BaSO_4$. But consider the new ion activities of the individual ions:

$$a_{Ba^{2+}} = 0.5 \times 10^{-5}$$

$$a_{Ca^{2+}} = 0.5 \times 10^{-3}$$

$$a_{SO_4^{2-}} = \underset{\substack{\text{contribution} \\ \text{from } BaSO_4}}{0.5 \times 10^{-5}} + \underset{\substack{\text{contribution} \\ \text{from } CaSO_4}}{0.5 \times 10^{-3}}$$

$$= 0.505 \times 10^{-3}$$

Because of the contribution from the $CaSO_4$ solution, the SO_4^{2-} activity is now much higher (fifty times) than it was in the $BaSO_4$ solution. Calculating the ion activity product for $BaSO_4$ in the mixed solution gives:

$$a_{Ba^{2+}} \cdot a_{SO_4^{2-}} = 0.5 \times 10^{-5} \times 0.505 \times 10^{-3}$$

$$= 10^{-8.6}$$

This is considerably greater than the solubility product of $BaSO_4$ at 25 °C. In spite of the two-fold dilution of Ba, the additional concentration of sulphate ions has made the solution supersaturated with $BaSO_4$, and we can expect precipitation of $BaSO_4$ to occur until the activity product has been reduced to the equilibrium value of 10^{-10}.

This unexpected outcome has arisen because $BaSO_4$ and $CaSO_4$ share an ionic species in common, the sulphate ion SO_4^{2-}. Had the second solution (b) consisted of calcium chloride $CaCl_2$, not $CaSO_4$, no extra sulphate ions would have been introduced, and no $BaSO_4$ would have precipitated. The precipitation of $BaSO_4$ by the addition of $CaSO_4$ is an example of the **common-ion effect**. The same outcome could be obtained by adding $BaCl_2$ solution instead of $CaSO_4$ (Ba^{2+} now being the common ion).

Barite deposits are found on the sea floor at places where barium-containing hydrothermal fluids (in which sulphur is present only as sulphide species like S^{2-}, H_2S and HS^-, and not as sulphate) emerge into sulphate-bearing sea water, a natural example of precipitation due to the common-ion effect.

Note that the simple quantitative concept of solubility applicable to single-salt solutions ceases to have any meaning in natural waters, where the activity of each ionic species can include contributions from a number of dissolved salts. This stresses the value of expressing solubility in the form of an equilibrium constant, the solubility product K. To summarize:

Ion activity product $> K$ supersaturated solution;
$\qquad = K$ saturated solution in equilibrium with solid phase;
$\qquad < K$ undersaturated solution.

4.2.2 Other kinds of equilibrium constant

SOLUBILITY OF A GAS

Water can dissolve gases as well as solids. A geologically important example is carbon dioxide (CO_2), which on dissolving forms a weak acid called **carbonic acid** (H_2CO_3). Water in equilibrium with atmospheric carbon dioxide is therefore always slightly acidic, a property relevant to many chemical weathering processes.

The dissolving of CO_2 in water can be written as a chemical reaction:

$$CO_2 + H_2O \rightleftharpoons H_2CO_3$$
$$\text{\small gas} \qquad \text{\small solution} \qquad \text{\small solution} \tag{4.14}$$

Using the appropriate ways of recording composition in these phases, the equilibrium constant is:

$$K_{CO_2} = a_{H_2CO_3} / P_{CO_2} \cdot X_{H_2O}$$
$$= a_{H_2CO_3} / P_{CO_2} \tag{4.15}$$

(since the mole fraction of water X_{H_2O} in dilute solution is very close to 1.00). P_{CO_2} is the partial pressure of CO_2, which in normal air is equal to 0.0003 atmospheres ($=30\,Pa$).

This equilibrium constant, though relating to the solubility of a solute species, is quite different in form from the solubility product of Equation 4.11. The reason is that the behaviour of CO_2 in a solution, as Equation 4.14 shows, is unlike that of ionic compounds, and this is reflected in the mathematical form of an equilibrium constant.

Increasing the air pressure or the proportion of CO_2 in air shifts equilibrium 4.14 to the right, increasing the solubility of CO_2 in water, as expressed by $a_{H_2CO_3}$. Opening a can of fizzy drink, on the other hand, releases the gas pressure inside, and as the dissolved CO_2 becomes supersaturated it appears as small bubbles of the gas phase, a process similar to the formation of vesicles in a molten lava (Chapter 2).

A gas like CO_2 is less soluble in hot water than cold (Box 4.2).

DISSOCIATION OF WEAK ACIDS

Carbonic acid is an example of a weak acid (Appendix C). Unlike the familiar strong acids such as HCl (hydrochloric acid), it dissociates into ions only to a small extent. This happens in two stages:

$$(1) \quad H_2CO_3 \rightleftharpoons H^+ + HCO_3^- \quad K_1 = \frac{a_{H^+} \cdot a_{HCO_3^-}}{a_{H_2CO_3}} = 10^{-6.4} \quad (4.16)$$

$$(2) \quad HCO_3^- \rightleftharpoons H^+ + CO_3^{2-} \quad K_2 = \frac{a_{H^+} \cdot a_{CO_3^{2-}}}{a_{HCO_3^-}} = 10^{-10.3} (4.17)$$

It is these dissociation reactions that make aqueous solutions of CO_2 acidic.

K_1 and K_2 represent a class of equilibrium constants known as **dissociation constants**. One finds with polybasic acids like H_2CO_3 (Appendix C) that K_1 is very much greater than K_2. The acidity of the solution is therefore almost entirely due to reaction 4.16 alone. Knowing the equilibrium constants for reactions 4.14 and 4.16, it is possible to calculate the pH (acidity) of water that has equilibrated with atmospheric carbon dioxide (Exercise 3).

The carbonic acid in natural waters determines whether they will dissolve carbonates (limestone) or precipitate them:

$$CaCO_3 + H_2CO_3 \underset{deposition}{\overset{weathering}{\rightleftharpoons}} Ca^{2+} + 2HCO_3^- \quad (4.18)$$

solid   solution   bicarbonate
solution

Physical conditions influence this equilibrium chiefly through the amount of dissolved CO_2 (as H_2CO_3). A local increase in atmospheric CO_2 content (owing to the decay of organic matter, for example), or an increase in total pressure leads to a higher H_2CO_3 concentration which shifts the equilibrium to the right (more $CaCO_3$ is dissolved). An increase in temperature, on the other hand, expels dissolved CO_2 and makes $CaCO_3$ less soluble (Box 4.2).

Warm surface ocean water is saturated or oversaturated with carbonate. The solubility of $CaCO_3$ increases with depth, however, partly because deep ocean water is colder, and one can identify a **carbonate compensation depth** or **lysocline** below which ocean water is undersaturated with carbonate. This change is encountered about 3 km down in the Pacific and about 4.5 km down in the Atlantic. It has been estimated that 80% or more of the calcareous shell material precipitated near the surface is redissolved during or after settling to the deep ocean floor.

4.3   Non-ideal solutions; activity coefficient

All of the solutes considered so far have been only slightly soluble in water (their solubility products have been very small numbers). The discussion has in effect been limited to very **dilute** solutions, in which the ions are so dispersed that electrostatic interactions between them ('ion–ion interactions') can be ignored. The behaviour of ionic species in such solutions can be accurately expressed, as we have seen, in terms of equilibrium constants that involve only the concentrations of the species of interest. Solutions sufficiently dilute to comply with this simple model of behaviour are called **ideal solutions**.

Geological reality is less simple, however. Most natural waters are complex, multi-salt solutions, whose properties may be far from ideal. Our attention may be on species like $BaSO_4$ that occur only in low concentrations, but the solutions in which they are dissolved will typically contain larger amounts of other more soluble salts such as chlorides or bicarbonates. In this stronger solution all ions present, including the dispersed Ba^{2+} and SO_4^{2-} ions, will experience many ion–ion interactions, which impede their freedom to react. In particular, Ba^{2+} and SO_4^{2-} ions, tangled up electrostatically with other types of ion, will be less likely to react with each other than if dissolved at the same concentration in pure water. This depression of reactivity is a function of the **total salt content** of the solution. When formulating equilibria between specific ionic species in such **non-ideal solutions**, account must therefore be taken of the concentrations of all solutes present, not just the species of interest.

Consider the solubility product of $BaSO_4$, measured (a) in pure water and (b) in 0.1 molal NaCl solution.

$$K_{BaSO_4} = a_{Ba^{2+}} \cdot a_{SO_4^{2-}} = \frac{m_{Ba^{2+}}}{m^{\ominus}} \cdot \frac{m_{SO_4^{2-}}}{m^{\ominus}} \tag{4.19}$$

(a) pure water: solubility product $= 1.0 \times 10^{-10}$ $\hspace{2em}$ (4.20a)

(b) NaCl solution: solubility product $= 7.5 \times 10^{-10}$ $\hspace{2em}$ (4.20b)

Notice that substantially more $BaSO_4$ will dissolve in the non-ideal saline solution (b) than in the same amount of pure water. The reason is that the reverse reaction responsible for restricting solubility –

$$\underset{\text{solid}}{BaSO_4} \leftarrow \underset{\text{solution}}{Ba^{2+} + SO_4^{2-}}$$

– is inhibited by the non-ideal, ion–ion interactions that Ba^{2+} and SO_4^{2-} ions experience in the presence of abundant Na^+ and Cl^- ions.

To be dealing with equilibrium 'constants' that vary according to the nature of the host solution is unacceptable. The problem can be overcome by re-defining activity so that it serves as a measure of 'effective concentration', incorporating the reduction of ionic reactivity in stronger solutions. A complete definition of activity is therefore:

$$a_{Ba^{2+}} = \gamma_{Ba^{2+}} \cdot \frac{m_{Ba^{2+}}}{m^{\ominus}}$$

$$a_{SO_4^{2-}} = \gamma_{SO_4^{2-}} \cdot \frac{m_{SO_4^{2-}}}{m^{\ominus}} \tag{4.21}$$

γ is the Greek letter gamma, and the functions $\gamma_{Ba^{2+}}$ and $\gamma_{SO_4^{2-}}$ are called the **activity coeffieients** of Ba^{2+} and SO_4^{2-} in the saline solution. An activity coefficient is simply a variable factor which expresses the degree of non-ideality of the solution (for ideal solutions it equals 1.00). Though it may look like a 'fiddle factor', the activity coefficient has some foundation in theory, and its value can in many cases be calculated with acceptable accuracy.

Expressing the equilibrium constant in terms of the newly defined activities:

$$\begin{aligned} K_{BaSO_4} &= a_{Ba^{2+}} \cdot a_{SO_4^{2-}} \\ &= \gamma_{Ba^{2+}} \cdot \gamma_{SO_4^{2-}} \left(\frac{m_{Ba^{2+}}}{m^{\ominus}} \cdot \frac{m_{SO_4^{2-}}}{m^{\ominus}} \right) \\ &= 10^{-10} \end{aligned} \tag{4.22}$$

This equation remains true for all circumstances. The solubility product (which by definition relates to the observed concentrations, $m_i/m^{\ominus}$, not effective concentrations a_i) is the quantity in brackets above, and this can vary in non-ideal solutions according to the values of the activity coefficients.

4.3.1   Ionic strength

The activity coefficient γ_i for a particular species i depends on the concentrations of all the solutes present in the solution. How is this overall 'strength' of the solution to be expressed?

Because departures from ideality arise from electrostatic interactions between ions, it is logical to devise a parameter that combines the amount of each type of ion present in solution and the charge on the ion. This is accomplished by the **ionic strength**, introduced by G. N. Lewis and M. Randall in 1921. The ionic strength I of a solution is given by the formula

$$I = \tfrac{1}{2} \sum_i m_i z_i^2 \tag{4.23}$$

Different values of the integer subscript i (1, 2, 3, etc.) identify in turn the various ionic species of the solution. m_i is the molality of species i (which can be established from a chemical analysis of the solution) and

z_i is the charge on the ion concerned divided by the charge on an electron. The summation symbol $\sum\limits_{i}$ adds together the $m_i z_i^2$ terms for all the values of i (i.e. for every species in solution).

Consider a solution in which NaCl is present at a molality of $0.1\,mol\,kg^{-1}$ and BaF_2 has a molality of $0.005\,mol\,kg^{-1}$ (less than saturation). The ionic strength I of this solution is given by:

$$
\begin{aligned}
I = \tfrac{1}{2}\{\ & 0.1 \times 1^2 && m_{Na^+} \cdot (z_{Na^+})^2 \\
& + 0.1 \times 1^2 && m_{Cl^-} \cdot (z_{Cl^-})^2 \\
& + 0.005 \times 2^2 && m_{Ba^{2+}} \cdot (z_{Ba^{2+}})^2 \\
& + (2 \times 0.005) \times 1^2\} && m_{F^-} \cdot (z_{F^-})^2 \\
= \ & 0.115\,mol\,kg^{-1}
\end{aligned}
$$

Why does the charge of each ion z_i appear in this formula as z_i^2? A rigorous explanation would require a digression into electrostatic field theory (Atkins, 1994), but we can see that the appearance of z_i^2 is plausible in the following way. The force attracting an ion to one of its oppositely charged neighbours is proportional to z_i. If $z_i = 2$, the ion will be twice as strongly attracted at a given distance as it would be if $z_i = 1$. But the stronger attraction will tend to draw ion i closer to the neighbour, increasing the attraction still further, making it more than twice as strong as that felt by a singly charged ion. This relationship is represented more faithfully by $m_i z_i^2$ than $m_i z_i$.

Natural waters span a considerable range of ionic strength, as these representative figures show:

	$I/mol\,kg^{-1}$
River water	<0.01
Sea water	0.7
Brines	1 – 10

4.4   Natural waters

There is no universal theoretical treatment capable of predicting non-ideal behaviour across the whole range of ionic strengths given above. The degree of ionic interaction and its influence on the properties of a solution change considerably as the ionic strength increases. It will be helpful to divide the spectrum of natural waters into smaller ranges of ionic strength, for which different assumptions and approximations apply.

4.4.1 River water ($I < 0.01\,mol\,kg^{-1}$); Debye–Hückel Theory

Table 4.2 shows the principal dissolved constituents of average river water. The reader can confirm that its ionic strength is about 0.002 $mol\,kg^{-1}$. Solutions as dilute as this exhibit the weakest ionic interactions, because the ions are widely separated from each other. There is nonetheless a tendency for each ion to attract a diffuse, continually changing jumble of oppositely charged ions around itself, aptly described by chemists as the **ionic atmosphere** of the ion concerned. Around each cation there is a slight statistical preponderance of anions, and vice versa. Like the phenomenon of hydration (Box 4.1), this tenuous ion–ion association is sufficient to depress the free energy of the ions in solution, making them less likely to take part in chemical reactions such as precipitation. The extent of this non-ideality can be estimated using a simple equation derived by P. J. W. Debye and E. Hückel in 1923, from a consideration of the free-energy change associated with the electrostatic properties of the ionic atmosphere:

$$\log_{10} \gamma_i = -A z_i^2 I^{\frac{1}{2}} \tag{4.24}$$

where:

γ_i = activity coefficient of ionic species i;
z_i = charge on ion i (1, 2, 3, etc.);
A = a constant characteristic of the solvent ($A = 0.509\,kg^{\frac{1}{2}}mol^{-\frac{1}{2}}$ for water at 25 °C);
I = ionic strength of the solution.

This equation is known as the **Debye–Hückel Equation**, which works accurately for non-ideal solutions of ionic strength up to $0.01\,mol\,kg^{-1}$.

Table 4.2 Composition of average river water.

Ion	Concentration (ppm = $mg\,kg^{-1}$)	Molality m_i ($10^{-3}\,mol\,kg^{-1}$)
HCO_3^-	58.3	0.955
Ca^{2+}	15.0	0.375
Na^+	4.1	0.274
Cl^-	7.8	0.220
Mg^{2+}	4.1	0.168
SO_4^{2-}	11.2	0.117
K^+	2.3	0.059

Fortunately most fresh waters fall into this category (broadly $I <$ $0.01 \, mol \, kg^{-1}$).

For univalent ions in Table 4.2:

$$A = 0.509 \, kg^{\frac{1}{2}} mol^{-\frac{1}{2}}$$

$$z_i = \pm 1, \quad \text{therefore} \; z_i^2 = 1$$

$$I = 0.0021 \, mol \, kg^{-1}$$

Therefore $\qquad I^{\frac{1}{2}} = 0.046 \, mol^{\frac{1}{2}} kg^{-\frac{1}{2}}$

Thus $\qquad\qquad \log \gamma_i = -0.0234$

and $\qquad\qquad\quad \gamma_i = 0.95$

Thus in river water the behaviour of univalent ions (Na^+, K^+, HCO_3^- and Cl^-) is only marginally non-ideal:

$$a_i = 0.95 \, m_i/m^{\ominus}$$

and assuming ideal behaviour would introduce an error of only 5% for each of these ions.

The appearance of z_i^2 in Equation 4.24 (for reasons similar to those following Equation 4.23) suggests that divalent ions will show a much larger departure from ideality. For $z_i^2 = 4$

$$\log \gamma_i = -0.0937$$

and $\qquad\qquad\quad \gamma_i = 0.81$

The activity of each divalent ion will therefore be about 20% below its molality. Even at an ionic strength as low as $0.002 \, mol \, kg^{-1}$, river water is perceptibly non-ideal. For example, from Equations 4.24 and 4.13 we would expect that sparingly soluble species like $BaSO_4$ would be about 25% more soluble than in pure water.

In addition to dissolved material and sediment, rivers transport the products of weathering and erosion in the form of colloidal suspension (Box 4.3).

4.4.2 Sea water ($I = 0.7 \, mol \, kg^{-1}$)

Sea water, as the principal medium of sediment deposition and the ultimate sink for the dissolved products of erosion and anthropogenic pollution, is geologically the most important category of natural water. Analyses show that it has remarkably constant composition across the world. Confining attention to the open oceans, both the salinity (the total salt content) and the concentration ratios between elements vary by less than 1%. In enclosed basins the composition of sea water may vary more widely owing to evaporation or fresh water runoff.

Box 4.3

Colloids

Colloids consist of ultrafine particles (usually much smaller than 1 micrometre) of one phase dispersed metastably in another. Most colloids fall into one of three categories:

Sol: Solid particles dispersed in a liquid, like the clay particles suspended in a river. Certain types of sol 'set' into a turbid, semi-solid form called a **gel** (e.g. gelatin).

Emulsion: One liquid dispersed in another (e.g. milk).

Aerosol: Liquid or solid particles dispersed in a gas (smoke, fog).

Because of their enormous surface area, the chemistry of colloids is dominated by the surface propeties of the colloidal particles. They possess a surface charge owing to the adhesion of ions. When dispersed in a solution of low ionic strength, inter-particle repulsion prevents coagulation into larger particles. In a strong aqueous solution, however, a dense ionic atmosphere forms around the particles; they repel each other less effectively; and they therefore aggregate into larger, more stable particles. This process, seen for example when lemon juice is added to milk, is called **flocculation**. Much of the silting that occurs in estuaries is due to flocculation of colloidal clay particles when the river water in which they are dispersed mixes with sea water.

Table 4.3 shows the global average composition of sea water. Calculation of the ionic strength, assuming all the constituents shown to be fully ionized, is $0.686 \, mol \, kg^{-1}$. This is well outside the range of composition to which Debye–Hückel theory is applicable.

The population of the ionic atmosphere around an ion in fresh water is essentially transient: ions are too dispersed for permanent associations between ions to be of any significance. In sea water, however, certain ions associate on a more permanent and specific basis. For example, Mg^{2+} and HCO_3^- (bicarbonate) ions are abundant enough for a significant proportion of them to combine to form the **ion pair** $MgHCO_3^+$.

Table 4.3 Principal ionic constituents of sea water.

Ion	Concentration (ppm = $mg \, kg^{-1}$)	Molality m_i ($10^{-3} \, mol \, kg^{-1}$)	% free ion (calc)*	γ_i measured[†]
Cl^-	19011	535.5	100	–
Na^+	10570	459.6	99	0.70
Mg^{2+}	1271	53.0	87	0.26
SO_4^{2-}	2664	27.8	54	0.07
Ca^{2+}	406	10.2	91	0.20
K^+	380	9.7	99	0.60
HCO_3^-	121	2.0	69	0.55
Br^-	66	0.8	–	–
CO_3^{2-}	18	0.3	9	0.02

* See Box 4.4. Percentage for Cl^- is assumed.
† See Berner (1971), Table 3.6.

About 19% of the bicarbonate in sea water is thought to be present in this **associated** from (Box 4.4). Thus, instead of considering just two ionic species

$$Mg^{2+} \qquad HCO_3^-$$

a realistic chemical model of sea water must distinguish three ionic species:

$$Mg^{2+} \qquad HCO_3^- \qquad MgHCO_3^+$$

as separate chemical entities. Ions can also associate by forming coordination **complexes** (Chapters 7 and 9) in which the cohesive force resembles a covalent bond rather than an ionic one. Complex formation is particularly prevalent among the transition metals.

The extent of ion pairing in sea water is shown in Box 4.4.

PH OF SEA WATER: CARBONATE EQUILIBRIA

In laboratory experiments, if we wish to adjust the pH of an aqueous solution, it is usual to add small amounts of a strong acid such as hydrochloric acid (HCl) or a strong alkali such as sodium hydroxide (NaOH). 'Strong' in this context means that the acid or base is completely ionized (Appendix C), and therefore a small addition delivers a large dose of H^+ or OH^- respectively to the solution.

Sea water, however, is devoid of strong acids and bases, and its pH is controlled instead by the dissociation behaviour of weak acids. The most abundant weak acid in the oceans is carbonic acid, whose partial dissociation into bicarbonate (HCO_3^-) and carbonate (CO_3^{2-}) ions has already been considered in Equations 4.16 and 4.17. Because of its capacity to limit the acidity of a solution by recombining with H^+ (the reverse reaction of Equation 4.16), the bicarbonate ion HCO_3^- is referred to as the **conjugate base** of carbonic acid (Appendix C). Equations 4.16 and 4.17 show that carbonic acid can undergo two stages of ionization and, as more than one conjugate base is involved (HCO_3^- and CO_3^{2-}), chemists refer to carbonic acid, perhaps slightly confusingly, as a **polybasic acid** (some chemists use the term **polyprotic** instead). A number of other weak acids that occur in sea water in small amounts, such as phosphoric acid (H_3PO_4), silicic aicd (H_2SiO_3) and boric acid (H_3BO_3), are also polybasic.

The reactions 4.14, 4.16 and 4.17 exert a strong regulating influence on the pH of the oceans. Imagine an experiment in which sufficient strong acid is added to a volume of sea water to increase a_{H+} tenfold (i.e. to reduce pH by one unit). Initially the product $a_{H+} \cdot a_{CO_3^{2-}}$ in Equation 4.17 will also rise by a factor of 10, but this represents a departure from equilibrium that will cause H^+ and CO_3^{2-} to react together to form HCO_3^- until $(a_{H+} \cdot a_{CO_3^{2-}})/a_{HCO_3^-}$ has been restored to $10^{-10.3}$. This reaction consumes some of the additional H^+, and will therefore shift the pH back toward its original value.

Box 4.4

The extent of ion-pairing in sea water

An ordinary chemical analysis of sea water, like that in Table 4.3, does not identify the actual species (ion pairs, complexes or free ions) in which the element occurs. For example, magnesium could be present in sea water in any of several alternative forms.

$$Mg^{2+} \quad MgSO_4^0 \quad MgCO_3^0 \quad MgHCO_3^+$$
$$\text{free ion} \quad \longleftarrow \text{ion pairs/complexes} \longrightarrow$$

(The superscript ^{0} signifies the zero charge on a neutral dissolved species.) Table 4.2 gives no clue as to how important each of these species might actually be, but clearly it must be true that:

$$m_{Mg}(\text{total}) = m_{Mg^{2+}} + m_{MgSO_4^0} + m_{MgCO_3^0} + m_{MgHCO_3^-}$$
$$= 53 \times 10^{-3} \, mol \, kg^{-1}$$

This is called a **mass balance equation**.

The stability of an ion pair or complex is measurable in terms of its **dissociation constant**. For example, the dissociation reaction

$$MgHCO_3^+ \rightleftharpoons Mg^{2+} + HCO_3^-$$

has a dissociation (equilibrium) constant given by:

$$K_{MgHCO_3^+} = \frac{a_{Mg^{2+}} \cdot a_{HCO_3^-}}{a_{MgHCO_3^+}}$$

According to laboratory experiment:

$$K_{MgHCO_3^+} = 10^{-1.16} \quad \text{at } 25\,°C$$

The pie charts show the molal proportions of principal cations and anions in sea water. The detached segments show the proportion of each ion involved in ion pairing (ignoring less significant pairings such as $CaHCO_3^+$). The remaining part of each chart represents the proportion left as free ions.

It is clear that a high proportion of certain anions like SO_4^{2-} are bound up in ion pairs. More than 40% of the sulphate ion in sea water appears to be paired, more or less equally, with Na^+ and Mg^{2+}, and only 55% of the sulphate present exists as free ions. No more than 10% of the carbonate ion appears to be present as free ions. As one would expect, pairing is more prevalent among divalent ions.

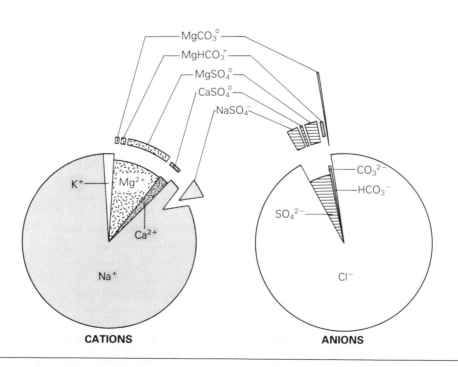

CATIONS   **ANIONS**

The higher level of H^+ will have the same effect on reaction 4.16: for the first instant $(a_{H^+} \cdot a_{HCO_3^-})/a_{H_2CO_3}$ will exceed $10^{-6.4}$ by a factor of ten, causing the added H^+ to react with HCO_3^- to form H_2CO_3. To use chemical parlance, equilibria 4.17 and 4.16 will both 'shift to the left' in response to the addition of acid, and in doing so consume a proportion of the added H^+. The net effect on the carbonate equilibria is to create more H_2CO_3 at the expense of CO_3^{2-}. (Had we added alkali instead of acid, the adjustment among carbonate species would have favoured CO_3^{2-} at the expense of H_2CO_3.) The *new* a_{H+} remains higher than the original value, but is by no means as high as it would have become had CO_3^{2-} and HCO_3^- not been present. This capacity of the carbonate system in the oceans to minimize the impact of changes in pH (in either direction) is called **buffering**. The combined buffering effect of carbonate, borate and phosphate systems serves to regulate the pH of sea water at a value close to 8.2 throughout the oceans of the world. Because the dissolved inorganic carbon (DIC) in the oceans represents a very large reservoir of carbonate, the capacity of this buffer system is huge.

The buffer mechanism would only cease to operate if the supply of dissolved CO_3^{2-} became exhausted. The activity of dissolved carbonate, however, is also regulated by the equilibrium:

$$CaCO_3 = Ca^{2+} + CO_3^{2-} \qquad K_{CaCO_3} = a_{Ca^{2+}} \cdot a_{CO_3^{2-}} = 10^{-8.3} \quad (4.25)$$

calcite   solution

where K_{CaCO_3} is the solubility product of calcite (Table 4.1). The oceans, at least at shallow depths, are teeming with suspended carbonate sediment and biogenic particles that can resupply dissolved CO_3^{2-} readily should it become depleted.

Many other chemical features of sea water are regulated by buffering reactions. If all of the Mg^{2+} delivered by rivers accumulated in the oceans, for example, sea water would be many times richer in Mg^{2+} than it actually is. The present level in the oceans is moderated by exchange reactions between sea water and ocean-floor basalts, which remove Mg^{2+} into minerals like chlorite that form as alteration products of the original ferromagnesian minerals in the basalt. It is important to recognize that the oceans do not simply accumulate all of the solute delivered by continental runoff; the concentrations of many elements in sea water are subject to complex regulatory mechanisms, in many of which biota play an important rôle.

4.4.3 Brines and hydrothermal fluids ($I > 1.0\,mol\,kg^{-1}$)

Near-surface ground waters in the continental crust are largely of **meteoric** origin; that is they are derived ultimately from atmospheric precipitation. They often have compositions of low ionic strength not very different from river water (Table 4.4), although depending to some extent on the type of rock through which they have flowed. In coastal areas there may also be a component of sea water present. In deeply buried sedimentary rocks, however, the pore waters are **connate** in origin, originating as sea water trapped during accumulation of the sediment. Drilling shows that such waters – oilfield brines, interstitial pore waters, 'formation water' and so on – are highly saline (Table 4.4), having remained in contact with the host lithologies at elevated temperatures for millions of years.

Hot hypersaline aqueous fluids are important in another context, the transport and emplacement of ore bodies. The compositions and temperatures of ore-forming 'hydrothermal' fluids can be established from the study under the microscope of the **fluid inclusions** left behind in the ore and gangue minerals (Box 4.5). It is clear that simple ionic solubilities of ore minerals (solubility products; Table 4.1) are many times too low to explain the dissolved-metal concentrations actually measured in such fluids, which are often in the 100–500 ppm range for metals like copper (Cu) and zinc (Zn). The discrepancy reflects the dramatic increase in the solubility of such metals brought about by **complexing** in highly saline hydrothermal fluids (Chapter 7, 'The coordinate bond', page 158).

Determining the chemical forms in which the metals are dissolved and transported in the hydrothermal fluids that circulate in the Earth's crust is an important step in understanding the formation of **hydrothermal ore deposits**. The complexes formed depend on the temperature and pH of the solution, and on the **ligands** present. The most

Table 4.4 Ground water and brine.

	Ground water in Mississippi sandstone at depth 40 m* (ppm)	Oilfield brine, Mississippi, at depth 3330 m[†] (ppm)
Cl^-	4.4	158 200
Na^+	60	59 500
K^+	4.1	538
Ca^{2+}	44	36 400
Mg^{2+}	11	1 730
Fe^{2+}	1.3	298
SO_4^{2-}	22	310
HCO_3^-	327	–
Zn	0	300

* From D. K. Todd (1980) *Groundwater hydrology*, 2nd edn, Wiley, New York (Table 7.3.)
[†] From H. L. Barnes (ed.) 1979. *Geochemistry of hydrothermal ore deposits*, 2nd edn, Wiley, New York (Table 1.1.)

Box 4.5

Fluid inclusions in minerals

During the growth of a crystal from a hydrothermal fluid, it is common for the growing lattice to enclose and trap a minute volume of fluid, accidentally preserving a small sample of it for subsequent microscopic examination. Such **fluid inclusions**, which vary in size from less than 1 μm to more than 100 μm, can be used to estimate the temperature and composition of the original fluid.

The post-enclosure development of a saline fluid inclusion is shown in the diagram, which resembles the phase diagram for pure water in Box 2.2 (p. 29). The saline fluid occupies a constant-volume enclosure (neglecting thermal contraction of the crystal) and therefore cooling causes the fluid pressure to fall from A to B along a constant-volume path called an **isochore**. At B the fluid becomes saturated with vapour, and after some supercooling a bubble nucleates at B_1. As the inclusion cools along the liquid/vapour phase boundary to room temperature (R), thermal contraction of the liquid allows the bubble to grow. The liquid may become saturated with one of its solutes (at D, say), so that a **daughter crystal** of halite or some other salt nucleates at D_1 and grows during subsequent cooling.

Using a microscope whose stage is specially equipped to heat the sample, the geologist can observe the cooling process in reverse, and the temperature at which the bubble just disappears indicates approximately the temperature at which the crystal and its inclusions were formed. The 'homogenization temperature' measured is actually T_B, but this is a useful minimum estimate of T_A. Measurement of the original pressure by other means allows T_A to be determined more accurately.

Like the salt used to melt ice on the roads in winter, the salinity of the fluid depresses the freezing point relative to that of pure water. The phase relations of the system water−salt resemble those of Di−An (Figure 2.4). Modern microscope 'heating stages' are also equipped to cool samples as far as −180 °C, making it possible to determine the temperature at which ice crystals first appear on cooling or disappear on warming up, from which the salinity of the fluid can be estimated.

Fluid inclusions provide a very powerful means of exploring the physical and chemical properties of hydrothermal systems. More information can be found in Rankin (1989) and the book by Shepherd *et al.* (1985).

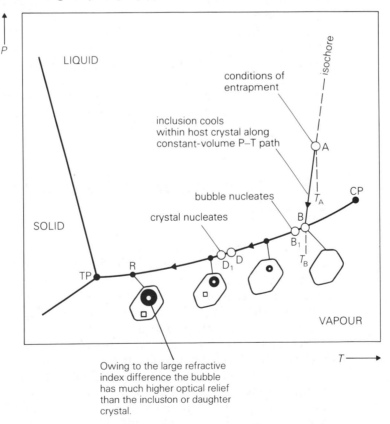

Owing to the large refractive index difference the bubble has much higher optical relief than the inclusion or daughter crystal.

important ligands (Appendix A) in typical ore-forming fluids are probably Cl^-, H_2S and HS^-. The sulphide ion itself (S^{2-}) is unlikely to be an important ligand in the neutral or mildly acidic fluids believed to be typical of ore-forming systems.

Because ore-forming fluids are known to be highly saline, **chloride complexes** dominate the hydrotheraml transport of many important metals. Thus lead (Pb) may be present as $PbCl^+$ or $PbCl_4^{2-}$, depending on the salinity:

$$Pb^{2+} + Cl^- \rightarrow PbCl^+ \qquad \text{low salinity}$$
$$Pb^{2+} + 4\,Cl^- \rightarrow PbCl_4^{2-} \quad \text{high salinity}$$

These species enable solutions in contact with galena (PbS) to contain as much as 600 parts per million of lead (by weight), whereas pure water saturated with PbS contains only 4×10^{-9} ppm.

Experiments and calculations suggest that the principal zinc species in hot saline fluids is neutral $ZnCl_2^0$, that silver is transported chiefly as $AgCl_2^-$, and tin as $SnCl^+$. Copper forms the complex $(CuCl)^0$ which dominates its hydrothermal solution chemistry above 250 °C, but at lower temperatures the bisulphide complex $Cu(HS)_3^{2-}$ may play a significant rôle.

The stability of such complexes is very sensitive to the temperature of the fluid, to its pH and to the salinity. It follows that if a sulphide-bearing hydrothermal fluid experiences a significant change in any of these variables, there may be a drastic reduction in the solubility of certain metals present, leading to their deposition as sulphide ore. This can happen if there is a drop in temperature (Box 4.2), mixing with other solutions of higher pH or lower salinity, or reaction with wall rocks.

4.5 Oxidation and reduction: *Eh*–pH diagrams

Oxidizing power and acidity are the two most important parameters of a sedimentary environment, jointly determining the limits of stability of all minerals that are found there. They are expressed in terms of the **oxidation potential** (*Eh*; Box 4.6) and the pH (Appendix C) of the solution with which the minerals coexist. Where several alternative minerals can crystallize depending on the conditions, it is logical to depict their stability fields on *Eh*–pH diagrams. As an illustration, Figure 4.1a shows the stability fields of various copper minerals in the presence of water, carbon dioxide, chloride and dissolved sulphur. The window shows the ranges of *Eh* and pH normally encountered in near-surface waters. The top and bottom of the window represent the limits beyond which water itself is unstable: beyond the top line the con-

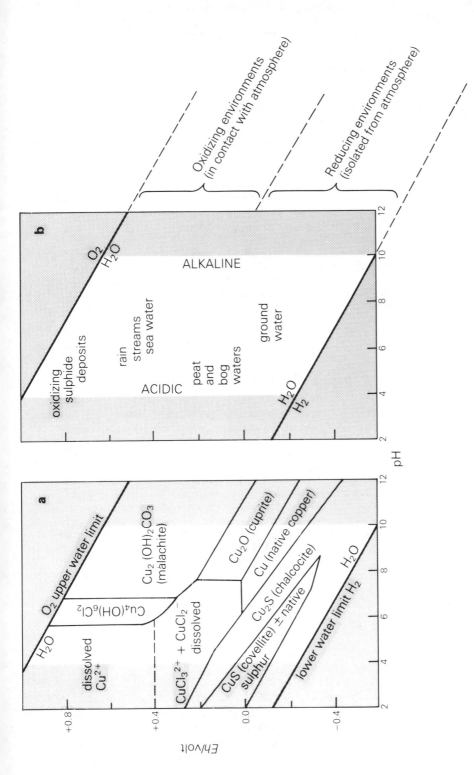

Figure 4.1 *Eh–pH* diagrams, showing a. approximate stability fields of copper minerals in equilibrium with water, S, Cl⁻ and CO_2 (simplified from Rose (1976) *Econ. Geol.*, **71**, with malachite data from Garrels and Christ (1965), after J. Anderson (1958)). b. approximate *Eh–pH* ranges of some natural aquatic environments: the stippled margins denote conditions beyond those normally encountered in near-surface waters (Krauskopf, 1979).

Box 4.6

Oxidation and reduction ('redox') reactions; *Eh*

In forming stable compounds with other elements, iron may take either of two **oxidation states**. A **ferrous** atom has donated or shared two of its electrons in forming bonds with other atoms and is said to be **divalent**, or in oxidation state II. Silicate minerals containing ferrous ions are commonly grey or green in colour (e.g. olivine). Alternatively the iron atom can commit three electrons to bonding, making a **trivalent** species (**ferric** iron, oxidation state III). Many compounds of ferric iron have a red or orange colour (e.g. rust). These states of iron can be symbolized Fe(II) and Fe(III). The notation does not necessarily mean that iron forms **ions** Fe^{2+} and Fe^{3+}: this depends on the type of bonding, as explained in Chapter 7. Pure metallic iron is said to be in oxidation state 0, written Fe(0). Oxidation states of metals are discussed in more detail in Chapter 9.

A reaction that causes iron to **increase** the number of its electrons involved in bonds with other atoms is called an **oxidation** reaction, as it increases the oxidation state of iron. An example is the weathering of the iron olivine fayalite (a ferrous compound):

$$2Fe_2SiO_4 + O_2 + H_2O \rightarrow 2Fe_2O_3 + 2H_2SiO_4$$

| olivine | air | water | hematite (ferric oxide) | dissolved form of SiO_2 |

Reduction is the opposite process, for example when an oxidized element receives back some or all of its bonding electrons. This can be seen for instance where basaltic lavas come into contact with bituminous shales:

$$2FeMgSiO_4 + C \rightarrow 2Fe + 2MgSiO_3 + CO_2$$

| 'olivine' component in melt: Fe(II) | carbon content of shales | 'native' iron: Fe(0) | 'pyroxene' component in melt |

This reaction illustrates an extremely important geochemical property of organic matter: its capacity to act as a natural **reducing agent** in a wide variety of geological circumstances, for example during the diagensis of sediment on the ocean floor.

Oxidation and reduction ('redox') reactions are important for many other elements that exist in nature in more than one oxidation state. Carbon, for example, can exist as the pure element (C[0]), or combined with oxygen ('oxidized carbon', C[IV]). In organic compounds, however, it is combined with hydrogen, an element of lower electronegativity (Figure 6.3), and in this state (C[−IV]) it is referred to as 'reduced carbon'. Sulphur behaves in a rather similar fashion; depending on the environment, it may occur:

(a) in the elemental state S[0] as the 'native' sulphur often deposited around volcanic vents (see front cover);

(b) as 'oxidized sulphur' in sulphur dioxide, SO_2 (S[IV]), and the sulphate ion, SO_4^{2-} (S[−VI]);

(c) as 'reduced sulphur' in H_2S (hydrogen sulphide, S[−II]) and in metal sulphides like galena (PbS).

The question of an element's oxidation state is discussed again in Chapter 9.

The stability of minerals containing these oxidation states depends on the conditions to which they are exposed. Environments open to the atmosphere are intrinsically oxidizing, because oxygen is good at drawing electrons away from metal atoms, and such conditions stabilize oxidized minerals like sulphates and hematite (Fe_2O_3). Environments dominated by organic matter, or cut off from the atmosphere (below the water table, for example), tend to be reducing. These are conditions favourable to the formation of sulphides and other oxygen-poor minerals (Figure 4.1). The relative oxidizing or reducing character of a natural solution, affecting the stability of minerals that coexist with it, is expressed in terms of its **oxidation** (or **redox**) **potential**, *Eh* (expressed in volts). *Eh* can be measured by putting a platinum electrode into the solution and reading the voltage it develops relative to a reference electrode. High values (approaching +IIV) indicate oxidizing conditions, whereas low (or negative) values represent reducing environments (Figure 4.1b).

ditions are so strongly oxidizing as to be capable of liberating molecular oxygen from water; whereas under the reducing conditions below the window, water breaks down to yield molecular hydrogen. Such extreme conditions are obviously not relevant to the hydrosphere. Note that these and many other boundaries on the diagram slope down to the right. This is because many oxidation and reduction reactions involve either H^+ or OH^- ions, making the *Eh* of the reaction sensitive to pH.

It is useful to compare Figure 4.1a with Figure 4.1b, which shows the Eh–pH ranges of waters from a variety of near-surface environments. Waters that circulate in contact with the atmosphere and remain well aerated are oxidizing. Ferrous minerals coexisting with them are eventually oxidized to ferric minerals such as hematite or goethite (hydrated Fe_2O_3), and iron-bearing rocks like shales and sandstones therefore weather a reddish colour. Stagnant or water-logged environments, particularly those rich in organic matter, tend to be strongly reducing. Weathering in these conditions, found below the water table, produces grey or green surfaces characteristic of ferrous-iron minerals. Minerals like magnetite and the sulphides are stable in such environments.

Eh–pH diagrams serve the same purpose for the sedimentary geochemist as P–T or T–X phase diagrams (Chapter 2) do for hard-rock petrologists. Many examples are given in the book by Garrels and Christ (1965).

4.6 Bibliography

Atkins, P. W. (1994) *Physical chemistry*, 5th edn, Oxford University Press, Oxford.

Berner, R. A. (1971) *Principles of chemical sedimentology*, McGraw-Hill, New York.

Edmonds, J. M. and von Damm, K. (1983) Hot springs and the ocean floor. *Scientific American* **248**, 70–85.

Garrels, R. M. and Christ, C. L. (1965) *Solutions, minerals and equilibria*, Harper, New York.

Krauskopf, K. B. (1995) *Introduction to geochemistry*, 3rd edn, McGraw-Hill, New York.

Libes, S. M. (1992) *An introduction to marine biogeochemistry*, Wiley, New York.

Open University (1989) *Ocean chemistry and deep-sea sediments*, Pergamon, Oxford.

Rankin, A. (1989) The physical chemistry of fluid inclusions, *Geology Today*, 5, 21–4.

Richards, H. G. and Strens, M. R. (1985) Ocean-floor hot springs, *Science Progress*, **69**, 341–58.

Shepherd, T. J., Rankin, A. H. and Alderton, D. H. M. (1985) *A practical guide to fluid inclusion studies*, Blackie, London.

4.7 Exercises

1 Calculate the solubility (in $mol\,kg^{-1}$) of $BaSO_4$ at 25 °C in:

   (a) pure water;
   (b) water containing $10^3\,mol\,kg^{-1}$ dissolved $CaSO_4$.

2 Calculate how many grams of CaF_2 will dissolve in $1\,kg$ of pure water at $25\,°C$. (Solubility product of CaF_2 at $25\,°C = 10^{-10.4}$; $Ca = 40$; $F = 19$.)

3 Calculate the molality of carbonic acid in rain water that has equilibrated with air. What is the pH of this solution (considering only reaction 4.16)? How is the solubility of CO_2 influenced by increase in (a) temperature and (b) total pressure? ($P_{CO_2}^{air} = 0.0003$ atmospheres. Equilibrium constant (reaction 4.15) $= 0.031\,mol\,kg^{-1}$ atmosphere^{-1} at $25\,°C$.)

5
Electrons in atoms

5.1 Does a geologist need to understand atoms?

We turn now to examine the behaviour of matter on the much smaller scale of individual atoms and molecules, which are about 10^{-10} m in size (Box 5.1). The sub-microscopic atomic world seems at first sight to have little bearing on everyday geology, which people usually associate with events on a much grander scale, such as earthquakes and volcanic eruptions. However, as the next few chapters show, many important properties of geological materials can be related to the types of atoms of which they are made and the chemical bonds that hold these atoms together. The geometry of a lava flow, for instance, depends on the viscosity of the lava: basaltic lava of low viscosity tends to spread out (if erupted on land) to form tabular flows which may flow a long distance and become quite thin; volcanoes constructed chiefly of basalt flows, such as the Hawaiian islands, are therefore characterized by very gentle slopes. Dacite or rhyolite lava, on the other hand, has a much higher visocity and tends to form a bulbous, steep-sided lava dome directly above the vent. Lava viscosity is determined by the bonding structure linking individual atoms together, and as we shall see in Chapter 7 this varies according to the dominant types of atom present. For example, the atomic characteristics of silicon (Si) are such that melts rich in SiO_2 (dacite and rhyolite, 65–75% SiO_2) are much stickier than silica-poor lavas (basalt, 45–52% SiO_2). Atomic interactions on the sub-nanometre scale thus exert a direct influence on the shape of geological structures some 10^{13} times larger.

The properties that geological materials inherit from their atomic constitution also have important economic applications. One geophysical approach to mineral exploration is to measure the electrical conductivity of the ground at a series of points in a promising area. The

---

Box 5.1

Units of atomic size

The size of an atom is measured in nanometres (nm): 1 nm $= 10^{-9}$ m or one thousand-millionth of a metre. Most atoms and ions have diameters in the range 0.1–0.3 nm.

The nanometre is the currently recognized SI unit for atomic size (Appendix B), but until quite recently the routine unit for atomic dimensions was the 'Ångstrom unit' (1 Å $= 0.1$ nm $= 10^{-10}$ m). Such units are still widely referred to in the literature.

---

chemical bonding in many sulphides makes them behave in some respects like metals, giving them a greater capacity to conduct electricity than the silicate rocks which host them (Chapter 7). This property can be exploited by the geophysicist to track down undiscovered ore-bodies. Once again we see the important influence of atomic characteristics on the macroscopic properties of minerals and other geological materials. A basic grasp of the nature of atoms and chemical bonding is therefore directly relevant to the work of the ordinary geologist.

5.2 The atom

Every atom consists of two components:

(a) a nucleus at the centre, containing nearly all of the mass of the atom but accounting for only one ten-thousandth of its diameter;
(b) a family of electrons gathered around the nucleus, forming a three-dimensional 'cloud' that makes up the volume of the atom.

The basic facts about these two atomic components are given in Table 5.1. The nucleus represents an extraordinarily dense state of matter, and it accommodates the atom's positive charge (which is proportional to Z, the number of protons it contains). Protons in such close proximity exert a powerful electrostatic repulsion on each other, but the nucleus is held together by a still more powerful force called the **Strong Nuclear Interaction** (Box 10.2).

In geochemistry we are more interested in the negatively charged **electrons** gathered around the nucleus, trapped in the electrostatic pull of its positive charge. The number of electrons is equal to the number of protons in the nucleus. These captive electrons provide the means by which atoms can associate and bond together. They are the currency of chemical reactions, being exchanged or shared whenever atoms interact and form bonds. It is therefore to the behaviour of electrons in atoms that the present chapter will be devoted.

Those reading this chapter for the first time may find that Box 5.4 provides a useful short cut.

Table 5.1 Basic facts about the atom.

	Nucleus	Electron cloud
Approximate size	10^{-14} m	10^{-10} m
Electrostatic charge	positive	negative
Constituent particles	protons and neutrons*	electrons
Approximate mass of individual particles	1.7×10^{-27} kg (about 1800 electron masses)	9×10^{-31} kg
Relative numbers of particles	in most types of nucleus, there are slightly more neutrons than protons	the number of electrons is equal to Z, the number of protons
Approximate density of matter ($kg\,dm^{-3} = g\,cm^{-3}$)	10^{12} kg dm^{-3}	1 kg dm^{-3}

* Protons carry one unit of positive charge and neutrons are electrically neutral. Electrons each have one unit of negative charge. In a neutral atom, the number of electrons must equal Z, the number of protons. Z is called the **atomic number**.

5.2.1 The mechanics of atomic particles

Mechanics is the science of bodies in motion. It describes the motion of anything from billiard balls to space statellites and planets. Except for the 20th-century contributions of relativity and quantum mechanics, the basic rules of mechanics have been known for more than two centuries, since the days of Isaac Newton. His contribution was of such importance that one often refers to the classical mechanics of the macroscopic world as 'Newtonian mechanics'.

It is natural to expect the movement of electrons around the nucleus in an atom to conform to the principles of Newtonian mechanics, like the planets in their motion around the Sun. The attractive forces in these two cases, though different in kind – electrostatic and gravitational respectively – conform to the same inverse square 'law' (Appendix B), so that one would expect the mathematical analysis to be the same. Nevertheless, it has been clear for about sixty years that classical mechanics is not wholly applicable to atomic and sub-atomic phenomena. Other influences seem to be at work which, though having little obvious impact on the macroscopic world, are paramount in the physics of the atom.

Perhaps the most radical departure from everyday experience is the notion that atomic particles such as electrons possess some of the properties of waves, a suggestion first made by Louis de Broglie in 1924. (Readers unfamiliar with the physics of waves may find the introduction in Box 5.2 helpful.) De Broglie's idea was soon reinforced

Box 5.2

What is a wave?

A wave describes a periodic disturbance in the value of some physical parameter. When a ripple travels across the surface of an otherwise still pond, the physical parameter being disturbed is the elevation of the pond surface: a twig floating on the surface will be seen to bob up and down as the ripple passes by, indicating that the pond surface is periodically displaced from its equilibrium position. A wave is periodic in both time and space: the twig bobs up and down v times per second (we call v the frequency of the wave, measured in units of s^{-1}; v is the Greek letter 'nu'), but if we take a snapshot at one instant we will see that successive wave crests are spaced out at a constant distance from each other, known as the wavelength λ of the wave (measured in metres; λ is the Greek letter 'lambda').

The ripple on the pond surface is the easiest type of wave to visualize, because the disturbance affects the position of a visible feature, the pond surface. The concept of 'waves' recognized by physics has a much wider scope, however, because physical quantities other than position may undergo periodic oscillation. A good example is a sound wave, in which it is the pressure of the air that fluctuates as the wave passes by. These changes in air pressure generate a periodic pressure difference across the eardrum that we sense as sound; the more times the pressure oscillates per second (the greater the frequency), the higher the 'pitch' of the note that we perceive. In the absence of air there is no pressure to fluctuate, which is why sound cannot propagate through a vacuum.

An electromagnetic wave (e.g. light) can be visualized as a train of 'ripples' in the intensity of electric and magnetic fields. In a place that is remote from magnets and electrostatic charges, the average or equilibrium values of these fields will both be zero, but the passage of a light wave causes them each to oscillate between positive and negative values. Light waves are characterized by wavelength and frequency in the same way as sound waves, though the values are very different (Box 6.3).

The examples of waves so far considered have been **travelling waves** which propagate energy from one place to another (e.g. from the loudspeaker to the ear). A wave confined within an enclosure of some kind, however, behaves in a different way: when it is reflected from the walls of the enclosure, the interaction between forward and reflected waves makes the wave appear to stand still. This **stationary wave** is simplest to see in the one-dimensional example of a guitar string. An important property of a stationary wave is that it has a clearly defined wavelength determined by the dimensions of the enclosure (e.g. the length of the guitar string). This phenomenon is exploited in organ pipes (sound) and lasers (light).

by the experimental discovery that a beam of electrons can be **diffracted** by a crystal lattice.

Diffraction is a phenomenon peculiar to waves, which arises when any kind of wave encounters a periodic (regularly repeated) structure whose repeat-distance is similar in magnitude to the wavelength of the wave concerned (Box 5.3). In the case of a crystal, this periodic structure is the regular three-dimensional array of its component atoms. The waves scattered from different parts of this structure **interfere** with each other, generating a characteristic spatial pattern of high and low beam intensities called a **diffraction pattern**, which can be recorded on a photographic plate. (Mineralogists and crystallographers use diffraction, either of X-rays – electromagnetic waves – or electrons, to investigate the internal structure of minerals and other crystalline materials.) Classical physics offers no explanation for the diffraction of electrons if they behave solely as particles, and the phenomenon provides firm evidence that the movements of atomic particles are determined by a underlying wave-like property.

If an electron is to be considered as a wave phenomenon, how can its position and motion be specified? Figure 5.1 depicts a moving electron as a wave-pulse, travelling in this case along the x-axis but frozen for inspection at some instant t, like racehorses at a photo-finish. The electron's position at time t cannot be defined precisely, because the wave-pulse extends smoothly over a range of x-values. The range Δx represents a fundamental interval of uncertainty, within which we cannot pinpoint exactly where the electron lies, though we know it lurks there somewhere. This uncertainty interval, which exceeds the physical size of the electron, is not the result of any experimental error, but must be seen as a fundamental physical limitation of the concept of 'position' where waves are concerned. Heisenberg, in 1927, was the first to recognize this. He expressed it in a quantitative form called the **Uncertainty Principle**, but the details need not concern us.

In examining the atom, therefore, we cannot look upon electrons as minute planets, orbiting the nucleus with precisely determined coordinates and motion. The wave nature of the electron rules out this precise classical image, introducing in its place a view of the electron in which position is subject to a degree of uncertainty. Although, as we shall see, electrons occupy well-defined spatial domains around the nucleus, the precise manner in which each patrols its own territory is concealed from us by fundamental limits of physical perception, which

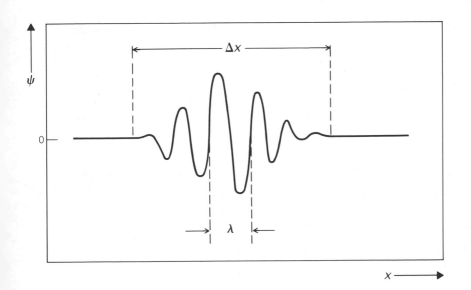

Figure 5.1 A wave train of restricted length, loosely illustrating the wave behaviour of an electron moving in free space. The horizontal axis is the x-direction along which the wave is travelling. The vertical axis represents the physical property (as yet unidentified) whose oscillation transmits the wave. It is called the *wave function*, and is symbolized by the Greek letter ψ (psi).

A familiar example of such a short wave 'pulse' is the wave observed when a stone is dropped into a still pond; viewed in cross-section, this would look very similar to the figure. In this case the wave function would represent the vertical displacement of the pond surface from its equilibrium (flat) state.

Box 5.3

Diffraction

Diffraction is a phenomenon that occurs when electromagnetic waves such as light interact with some regularly repeated geometrical pattern. A familiar example is provided by the colour fringes seen in light reflected from the surface of a compact disc. The key requirement for diffraction to occur is that the repeat distance of the pattern should be similar in magnitude to the wavelength(s) of the light being diffracted. The digital signal on a CD is etched in a spiral track with a constant spacing between successive turns; in a manufactured **diffraction grating** the same effect is achieved by engraving straight grooves on a glass plate or mirror (with a typical spacing of 500–1000 nm). In each case, the spacing of track or grooves happens (or is designed) to lie within the wavelength range of visible radiation (400–760 nm). For the same reason X-rays ($\lambda = 10^{-2}$ to 1 nm) are diffracted by the regularly repeated atomic structure of a crystal (typical repeat distance 0.1 to 2 nm), which acts as a 3D diffraction grating.

As a glance at any CD demonstrates, the effect of a diffraction grating on visible radiation is very similar to that of a prism: white light is spread out into a continuum of different colours. The physical process involved, however, is entirely different. Incoming waves are scattered in all directions by individual grooves or tracks. For rays scattered in certain specific directions, waves of a particular wavelength emanating from neighbouring grooves (or atoms in the case of X-ray diffraction) will reinforce each other and give an enhanced intensity of that wavelength (see figure), whereas in other directions neighbouring waves are 'out of phase' and will eliminate each other.

When many wavelengths are present in the incoming light beam, the effect is to disperse one wavelength in one specific angular direction but not in others, so that a spectrum of colours is formed.

Diffraction finds many uses in spectrometry (Chapter 6) for separating visible, ultraviolet or X-ray spectra (Box 6.3) into spectral 'lines' to allow the emissions of individual elements to be measured. The wavelength of the line can be calculated from the angle of diffraction at which the wavelength is detected and the spacing of the grooves or atomic planes. This relationship is expressed most simply for X-ray diffraction, through the well-known Bragg equation:

$$n\lambda = 2d\sin\theta$$

where n is a whole number (1, 2 ...), d is the spacing between adjacent planes of identical atoms in the crystal and θ is the diffraction angle at which the X-ray wavelength λ will be diffracted with maximum intensity. Using a crystal with known d-spacing, the Bragg equation can be used to investigate the various wavelengths present in a complex X-ray spectrum from a geological specimen (Box 6.4); on the other hand, if the X-ray spectrum consists of a single known wavelength from an X-ray tube, the equation provides a tool for investigating the atomic structure of an unknown crystalline material by measuring the d-spacings of various sets of atomic planes. Since the work of the Braggs in the 1930s, X-ray diffracton has been an essential technique for determining the atomic structure of minerals.

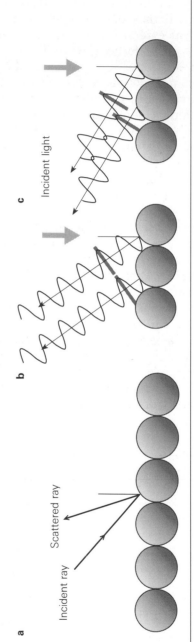

Principle of diffraction grating illustrated by a row of atoms. a. Incident ray, and ray scattered by atom at arbitrary angle. b. Waves scattered from neighbouring atoms such that they are **in phase** (constructive interference – rays pull together – enhanced ray diffracted in this direction). c. Waves scattered from neighbouring atoms such that they are **out of phase** (destructive interference – rays cancel each other out – zero intensity observed in this direction). Broad bars illustrate relative positions of wave-

must be embraced in any model of the atom that we develop. The mathematical certainty enshrined in the laws of Newtonian mechanics gives way, on the scale of the atom, to a statistical interpretation of particle mechanics based on probability.

5.3   Stationary waves

An electron that belongs to an atom is confined to a small volume of space close to the nucleus. How does a wave behave in these circumstances? To answer this question it is helpful to look at waves of a more familiar kind, those on a vibrating string.

When a guitar string is plucked, the lateral displacement y introduced by the player's finger initiates a rapid transverse vibration of the string, which is communicated as waves moving away towards each end of the string. On a string of infinite length, each of these disturbances would continue to propagate outwards indefinitely. The waves carry the energy of the disturbance away from the point of plucking, where the string will soon resume a stationary condition. A guitar string however has a restricted length, and waves reaching the fixed ends of the string are reflected back on themselves. The string is therefore deflected by waves travelling in opposite directions at the same time, and the overall effect is to set up a **stationary wave**: the string vibrates rapidly up and down within a stationary **envelope** that gradually contracts with time. This envelope on the guitar string is just visible to the naked eye. Its form is shown (exaggerated) at the top of Figure 5.2. The stationary wave is the characteristic form adopted by any wave that is trapped in a restricted region of space (like a guitar string, or an organ pipe). Standing waves are fundamental to the generation of sound, but the phenomenon is not restricted to acoustic waves.

When an electron is captured by an atom, by the electrostatic attraction of the the nucleus, the attendant wave becomes trapped within the volume of the atom. It responds to confinement in the same way as the guitar string: it becomes a stationary wave, an oscillating disturbance inside a fixed envelope.

Before developing this analogy further, one must recognize two obvious limitations:

(a)   On the guitar we see a wave distributed along a one-dimensional string, whereas the electron must be treated as a wave in three-dimensional space.

(b)   The vibration of a plucked string decays away quite rapidly because its energy is being dissipated into the surrounding air, through which the sound reaches our ears. The electron standing-wave does not experience this 'damping' effect, and continues vibrating.

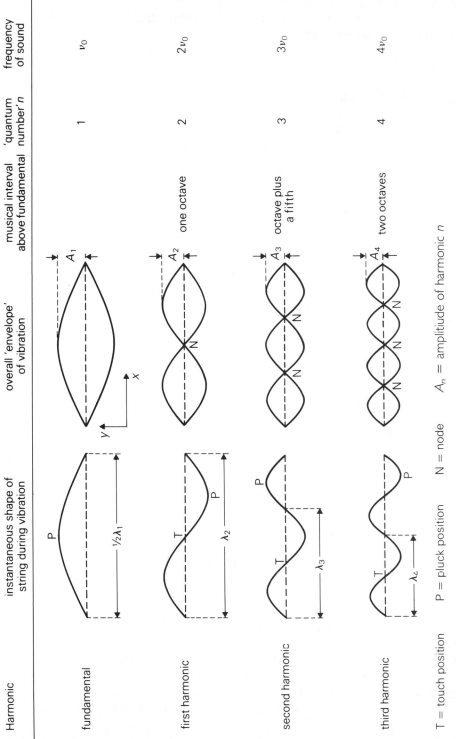

Harmonic	instantaneous shape of string during vibration	overall 'envelope' of vibration	musical interval above fundamental	'quantum number' n	frequency of sound
fundamental				1	ν_0
first harmonic			one octave	2	$2\nu_0$
second harmonic			octave plus a fifth	3	$3\nu_0$
third harmonic			two octaves	4	$4\nu_0$

T = touch position   P = pluck position   N = node A_n = amplitude of harmonic n

Figure 5.2 The harmonics of a plucked string. To obtain such harmonics on the guitar, the player touches the string lightly at the appropriate nodal point (identified as T), whilst plucking it sharply near to the other end (point P). The finger at T is removed immediately after plucking. The idea is to establish a nodal point at T while allowing the string to oscillate along its entire length.

The characteristics of these two types of standing waves differ in detail, but the underlying principles are identical.

5.3.1   Harmonics

The stationary wave shown at the top of Figure 5.2 is the simplest possible mode of vibration, obtained by plucking the string at its centre. This waveform, the **fundamental**, produces the lowest musical note obtainable from the string (frequency v_0). To produce higher notes from the same guitar string one can shorten its length by pressing it down, but it is more relevant to consider what other notes are obtainable on the open (full-length) string. It is possible to generate a number of higher notes from the open guitar string by making it adopt alternative forms of stationary waves called **harmonics**, as described in the caption to Figure 5.2. (Harmonics on the guitar have a characteristic ethereal tone, and are commonly used in contemporary guitar music to extend the instrument's range of tonal colour.)

Because the ends of the string are immobilized, only a restricted number of harmonics is possible. These all have wavelengths (λ_2, λ_3, etc.) and frequencies (v_2, v_3, etc.) that are related in simple ways to the fundamental as shown in Figure 5.2. Whereas in the fundamental mode the string is stationary only at its ends, the harmonics share the property of having intermediate points where the string remains stationary too (that is, where the lateral displacement of the string, y, remains zero). These points are called **nodes**, and are denoted by the symbol 'N' in Figure 5.2.

The fundamental and the harmonics are members of a restricted series of stationary waveforms that can reside on a string of length L. They can all be represented by one general equation:

$$y = A_n \sin\left(\frac{n}{2} \cdot \frac{x}{L} \cdot 360°\right) \tag{5.1}$$

where y represents the lateral displacement of the string (from its equilibrium position) at distance x from the end of the string. A_n measures the maximum displacement, and is called the **amplitude**. The integer n has a different value for each of the possible harmonics. Entering '$n = 1$' into Equation 5.1 yields the equation describing the fundamental; '$n = 2$' gives the equation for the first harmonic (Figure 5.2) and so on. Note that the value of n defines several important features of a harmonic:

(a) n indicates the number of maxima and minima (the number of 'lobes') of the waveform;
(b)   the number of nodes is equal to $n - 1$;
(c)   the wavelength of the stationary wave is equal to $2L/n$ (Figure 5.2);
(d)   the frequency of vibration of each harmonic is proportional to n.

113

The important conclusion is that a string fixed at both ends can accommodate only a restricted number of stationary waves compatible with its length L, so the vibration frequency can adopt only a limited number of values (the musical pitches of the fundamental and harmonics). A variable that is allowed by the properties of the system to have only certain discrete values is said to be **quantized**. The integer n which enumerates these permitted values is called a **quantum number**. These terms are used chiefly in the context of atomic physics, but they identify characteristics common to all types of stationary waves, regardless of scale.

5.4   Electron waves in atoms

The harmonics of a vibrating string can be investigated using simple apparatus, so a thorough mathematical treatment is not called for. Because of its size the atom does not lend itself to such simple experimentation, and our understanding of how it works comes mainly from theoretical physics. Erwin Schrödinger, in 1926, formulated an elegant and highly successful mathematical theory of particle mechanics incorporating the wave-like properties of the electron. His method of analysis is known as **wave mechanics**, and it provides the foundation of modern atomic (and nuclear) physics. Though the mathematics is difficult, the underlying physical concepts are straightforward, and have a lot in common with the analysis of the vibrating string.

Schrödinger set up a general **wave equation** describing the physical circumstances of the electron in an atom: the nature of the electrostatic force attracting it toward the nucleus (the inverse-square law of classical physics) and the newly recognized wave properties of the electron itself. Schrödinger's work suggested that the electron trapped in an atom behaves in much the same way as any stationary wave, including the one on a stringed instrument (Table 5.2).

The Schrödinger wave equation is a differential equation, which offers a number of possible mathematical 'solutions' called **stationary states**. Each simply describes a different stationary 'waveform' a trapped electron can adopt, analogous to those shown in Figure 5.2. Each distinct electronic 'waveform' with its own specific three-dimensional geometry is called an **orbital**. One speaks of an electron 'occupying' a particular orbital, reminding us that an orbital is the wave-mechanical equivalent of a planetary orbit.

Equation 5.1 expresses the way the displacement y varies along the length of the vibrating string (the x dimension). In the vocabulary of wave theory, the mathematical function y is called the **wave function**. Each solution of the Schrödinger equation expresses how a wave function ψ (the Greek letter 'psi') varies in three-dimensional space (x, y, z) around the nucleus. To understand the physical significance of ψ, we need to square it. It is true of most types of wave that the intensity of

Table 5.2 Parallels between a vibrating string and an electron orbital.

	Stationary wave on a guitar string	Electron in an atom
Nature of wave	Wave in *one dimension* (along string) Extent of wave restricted by fixed ends	Wave in *three dimensions* (around nucleus) Spatial extent of electron standing wave restricted by electrostatic pull of nucleus
Stationary states	*One* quantum number (n) sufficient to define stationary states: $n = 1$ fundamental $n = 2$ first harmonic $n = 3$ second harmonic, etc.	*Four* quantum numbers ($n, l, m, s*$) are required in order to define all possible electron stationary states; n and l are the most important:

	$l = 0$	$l = 1$	$l = 2$	$l = 3$
$n = 1$	1s	–	–	–
$n = 2$	2s	2p	–	–
$n = 3$	3s	3p	3d	–
$n = 4$	4s	4p	4d	4f

	Stationary wave on a guitar string	Electron in an atom
Significance of wave function	Wave function = y (lateral displacement of string) Perceived quantity: 'loudness' $\propto y^2$	Wave function = ψ Physical significance: ψ^2 = probability of finding the electron at coordinates (x, y, z). = 'electron density' at (x, y, z).

* n = principal quantum number
l = angular momentum quantum number
m = magnetic quantum number
s = spin quantum number   } See Table 5.3.

the physical seansation is proportional to the square of the wave function. For example, the loudness of the sound emanating from the guitar string is proportional to y^2, not y.

The magnitude of ψ^2 at each point in the atom tells us the probability of finding the electron at that point in space. In keeping with the Uncertainty Principle, this probability is less than one at every individual point (that is, there is no one point at which we can locate the electron with certainty), but considering the atom as a whole it must add up to one for each of the electrons present.

Solutions to the Schrödinger equation for the electron in a atom are independent of time (since, like Equation 5.1, they contain no t terms), and so they cannot indicate the precise trajectory or orbit followed by the electron, showing how x, y and z vary as a function of time. The electron must be visualized as a constant but diffuse cloud extending throughout the volume of the orbital, the **electron density** of the cloud varying from point to point according to the magnitude of ψ^2.

For the vibrating string, one quantum number n is sufficient to enumerate the various types of stationary wave observed. The electron wave in the three-dimensional atom presents a more complicated picture, and four quantum numbers (Table 5.3) are required to encompass all

Table 5.3 Physical significance of quantum numbers.

Quantum number		Permitted values	Influence on geometry of orbital
Name	Symbol		

Principal quantum number	n	integer	(a) Determines the *size* of the orbital:
			
			    low n: compact orbital
			    high n: spread-out orbital
			    (Figures 5.3 and 5.4)
			
			 (b) $(n-1)$ is the number of *nodal surfaces where* $\psi^2 = 0$

Angular momentum quantum number	l	integer 0 to $n-1$	Determines the *shape* of the orbital: $l = 0$'s-orbital': spherical symmetry $l = 1$ 'p-orbital': electron density forms two balloons on opposite sides of nucleus (Figure 5.4) $l = 2$ 'd-orbital': electron density forms *four* balloons (Figure 5.4) $l = 3$ 'f-orbital': still more complex
Magnetic quantum number	m	integer $-l$ to $+l$	Determines the *orientation* of the orbital, e.g., indicates whether a p-orbital is aligned along the x, y or z axis

Note: The spin quantum number (s) only becomes relevant when multi-electron atoms are considered. It has only two permitted values, $-\frac{1}{2}$ and $+\frac{1}{2}$.

possible stationary states that the electron wave can adopt. For most purposes, we shall only need to consider two of these four quantum numbers: n, the **principal quantum number**, and l, the **angular-momentum quantum number**. (The physical origins of these names need not concern the reader.) The significance of each quantum number, summarized in Table 5.3, will become clear presently.

5.5   The shapes of electron orbitals

Orbitals come in all shapes and sizes, and some acquaintance with their geometry will be needed in order to understand the shapes of molecules and the internal structures of crystals (Chapter 8). The internal structure of the mineral diamond, for example, is a direct expression of how electron density is arranged within each constituent carbon atom. In a similar way, the disposition of electron orbitals in the oxygen atom is responsible for the bent shape of the water molecule, upon which the unique solvent power of water depends (Box 4.1).

The **symmetry** of an orbital is determined by the value of the quantum number l, in the manner outlined in Table 5.3. Orbitals for which l is zero have simple spherical symmetry, but as l increases we encounter progressively more complex symmetry. We shall begin with the simplest type of orbital symmetry, in which the parallel with the vibrating string is most apparent.

5.5.1   s-Orbitals

The simplest solutions to the Schrödinger wave equation, for which l is zero, all possess spherical symmetry and therefore ψ and ψ^2 can be depicted simply in terms of a radial coordinate r, the distance from the nucleus. Such solutions are called **s-orbitals**. The two simplest cases are shown in Figure 5.3. The upper half shows what we may think of as the atomic counterpart of the fundamental mode on a vibrating string (Figure 5.2). In keeping with this interpretation, the principal quantum number n is equal to 1 and the number of nodes is zero. This orbital is designated '1s'. On the right-hand side of Figure 5.3 is an attempt to show what a cross-section of this orbital would look like. The electron density (represented by the density of dots) is greatest immediately around the nucleus, and decreases smoothly away from it with an exponential-like profile. This diffuse outer fringe is common to all types of orbital, and in this respect wave-mechanical waveforms differ significantly from vibrating string harmonics, which of course terminate abruptly at the end of the string. All atoms therefore have diffuse outer margins.

The lower part of Figure 5.3 illustrates a more complex spherical orbital, designated 2s, which resembles a first harmonic: n has the value 2, and there is a node-like feature where the wave function and therefore the electron density ψ^2 are both zero. In three dimensions, this is actually a spherical nodal surface, separating a core of electron density from an outer fringe. These two parts of the orbital together accommodate the same electron which, rather paradoxically, distributes itself statistically between them. Note that the electron density extends significantly further from the nucleus than was the case for the 1s orbital (Figure 5.3).

Each of these orbitals is uniquely defined by the values of the two quantum numbers n (1 or 2) and l (0). A series of progressively larger, and more complex, spherical orbitals exists corresponding to values of n up to about 7, and these are distinguished as 3s, 4s and so on according to the value of n.

5.5.2   p-Orbitals

The p-orbitals are a second class of solution (identified by l having the value 1), in which electron density is concentrated into two 'balloons' which stick out from the nucleus in opposite directions (Figure 5.4). The lack of spherical symmetry here requires the introduction of arbitrary x, y and z axes, centred on the nucleus, in order to specify the orientation of these balloons. The Schrödinger equation indicates that there are three equivalent, but independent, forms of p-orbital, designated p_x, p_y and p_z, in which these balloons are aligned along the x, y and z axes respectively (Figure 5.5). These variants share the same values of n and l, but are distinguished by having different values of the

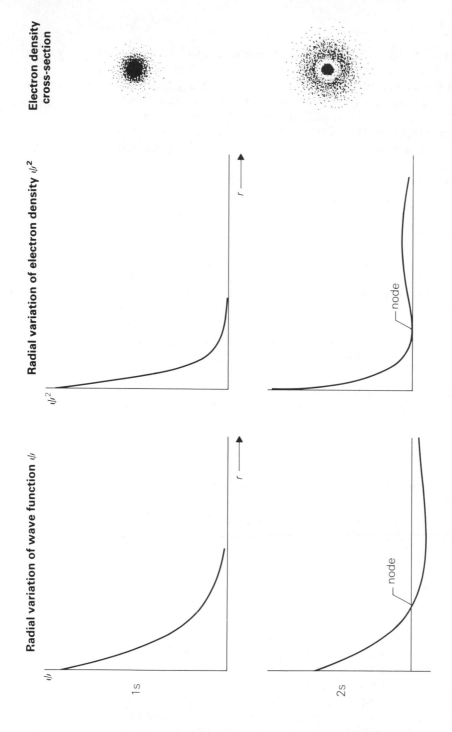

Figure 5.3 Alternative ways of visualizing s-orbitals. In the cross-section representations on the right, the density of dots indicates the electron density.

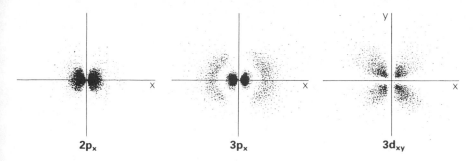

$2p_x$ $3p_x$ $3d_{xy}$

Figure 5.4 Cross-sections of electron density distributions in 2p, 3p and 3d orbitals. Electron density is represented by the density of dots.

magnetic quantum number m (Table 5.3), which serves to define the different **orientation** of each of these orbitals.

For reasons we need not discuss, the Schrödinger equation prohibits values of l greater than $n - 1$ (Table 5.3). Accordingly there is no p-orbital when $n = 1$, and the simplest p-orbitals ($2p_x$, $2p_y$ and $2p_z$) are encountered when $n = 2$. (Consistent with this value of n, we observe the existence of a nodal plane separating the two balloons, which has a similar significance to the nodal surface in the 2s orbital.) As with s-orbitals, the maximum electron density is found close to the nucleus (Figure 5.4). Three such p-orbitals exist for each value of n greater than 1, and these families are designated 2p, 3p, 4p, and so on according to the value n. (For most purposes, the subscripts x, y and z may be omitted.) As Figure 5.4 illustrates, the size of the orbital and the number of nodal surfaces increases with the value of n.

5.5.3 d-Orbitals

Setting l equal to 2 generates a third type of orbital called a d-orbital, which is first encountered when $n = 3$. Except for one special case, d-orbitals consist of four elongated balloons of electron density, extending out from the nucleus at right angles to each other (Figure 5.4). As stated in Table 5.3, m can now have the values -2, -1, 0, $+1$ and $+2$, and we therefore find five equivalent d-orbitals for each value of n (greater than 2). Their orientation in space is shown schematically in Figure 5.5. Three orbitals, d_{xy}, d_{yz} and d_{zx}, have balloons extending between the coordinate axes, whereas the other two have their electron density directed broadly along the axes. Our experience of wave theory so far suggests that, because $n = 3$, we should encounter two nodal surfaces in each of the 3d orbitals, and the nodal planes or surfaces evident in Figure 5.4 confirm this expectation. A separate family of five d-orbitals exists for every value of n greater than 2 ($= 3d$, 4d, and so on).

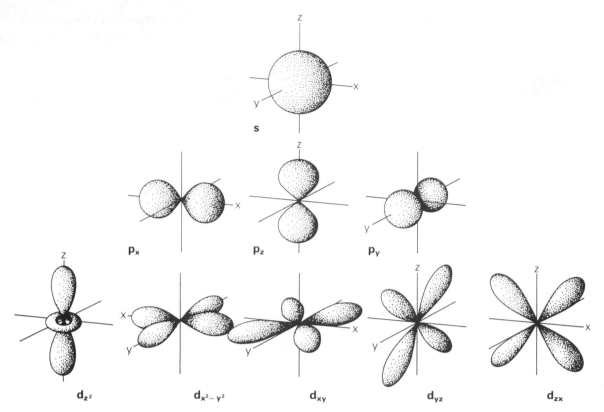

Figure 5.5 Simple 'balloon' diagrams showing where electron density is concentrated in s, p and d orbitals. Each balloon can be considered as a three-dimensional contour of electron density.

As we shall see in Chapter 9, d-orbitals are responsible for the distinctive chemistry of the transition metals such as iron, copper and gold, and play an important rôle in the aqueous-complex formation upon which their hydrothermal transport and deposition in ore bodies depend.

5.5.4 f-Orbitals

The final class of solutions to Schrödinger's equation are those for which $l = 3$. Their geometry is too complex to consider here, but we should note that they first occur when $n = 4$, and that seven equivalent orbitals having various orientations will exist for each value of n above 3. The f-orbitals only become important chemically in the context of heavy elements such as cerium and uranium.

5.6 Electron energy levels

Each stationary state of the Schrödinger equation has a well-defined value of the total electron energy E (the sum of the electron's potential and kinetic energies). The energy of an electron in an atom is therefore not free to vary continuously, but like the frequency of a guitar string is **quantized** and has to conform to one of these permitted **energy levels**. Niels Bohr had been the first to suspect this on empirical grounds in 1913, but the theoretical basis remained obscure until Schrödinger's work showed it to be a straightforward consequence of the stationary wave(s) set up by the trapped electron(s).

The energy levels of the various orbitals are shown in relation to each other (for the hydrogen atom) in Figure 5.6. The zero on the energy scale is defined as the energy of a 'free electron at rest', one that does not belong to any atom (see caption) and possesses no kinetic energy. This convention provides a common baseline that allows us to compare directly the energy levels of electrons in different types of atom.

Each box in Figure 5.6 represents an orbital. Considered together, the orbitals resemble an irregular set of 'pigeonholes' in energy space, offering the electron a variety of alternative accommodation. The 1s orbital has by far the lowest energy level, indicating that in this state the electron is most firmly bound to the nucleus. This interpretation is consistent with the very small size of the 1s orbital (Figure 5.3), which confines the electron very closely to the nucleus, where its electrostatic pull is strongest. This is the most stable stationary state the electron can adopt in the atom, the one in which its energy is minimized (Chapter 1). When the electron resides in this orbital, the hydrogen atom is said to be in its **ground state**.

The 2s and the three 2p orbitals share the same energy level, some distance up the energy scale. In spite of their different spatial configurations, therefore, they are energetically equivalent, at least in the hydrogen atom. The same is true of the one 3s orbital, the three 3p orbitals and the five 3d orbitals, all of which share a still higher energy level. Note that there is a direct relationship between an orbital's relative size and its energy level. Evidently an electron must possess quite a high energy before it can overcome the nuclear attraction sufficiently to spread itself into these more far-flung provinces of the atom's territory. In the hydrogen atom these orbitals are normally unoccupied, except temporarily (Chapter 6).

As n increases further, the energy levels get progressively closer together, and so the highest levels have been omitted from Figure 5.6 for clarity.

5.6.1 Multi-electron atoms

In order to describe the chemistry of elements other than hydrogen, the Schrödinger model needs to be extended to explain how more than one

Figure 5.6 A scale diagram of the energy-level structure of orbitals in the hydrogen atom (only one electron). Note that the 2s and 2p orbitals have the same energy, as do the 3s, 3p and 3d orbitals. Note also that the zero on the electron energy scale is equivalent to the energy of a free electron at rest. A negative electron energy therefore signifies that the electron is trapped in the atom by the nuclear field. The more negative the energy, the more tightly bound the electron has become, and the harder it is to remove from the atom. Positive energy values signify electrons having appreciable kinetic energy.

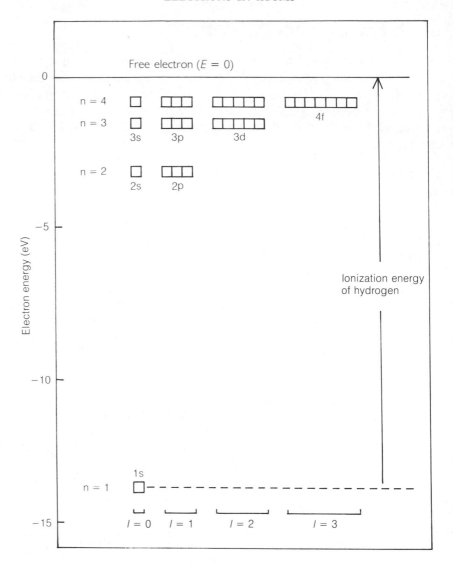

electron can be accommodated together in a single atom. Can separate electrons adopt identical waveforms in the same atom, or does wave mechanics organize them into different spatial domains around the nucleus?

The answer to this vital question was provided by Wolfgang Pauli, who, in 1925 (actually a year before Schrödinger's paper) formulated the **Exclusion Principle**. Stated in wave-mechanical terms, it says: 'No two electrons in the same atom may possess identical values of all four quantum numbers'. At this point the fourth quantum number s, representing the 'spin' of the electron, enters the discussion. According to the Exclusion Principle, two electrons in the same atom may possess the same values of n, l and m (that is, they may 'occupy the same orbital') only *if* they have different values of *s*. Wave mechanics allows *s* only two possible values, $-\frac{1}{2}$ and $+\frac{1}{2}$ ('$\frac{1}{2}$' is the contribution that electron spin makes to the atom's angular momentum, but the details need not concern us.) The important property of the spin is that it can have one of two directions, a positive or negative 'sense'. Although 'spin' in this quantized, wave-mechanical form is rather an abstract concept, it leads to the important practical conclusion that each of the orbitals we have been considering can accept two electrons, subject only to the proviso that their spins are opposed.

We can now predict the arrangement of electrons in any atom. Each electron entering an atom will of course occupy the orbital that offers accommodation at the lowest available energy level. In the atom of helium, the two electrons present can share the 1s orbital. However in the lithium atom with three electrons, the third electron cannot, according to the Exclusion Principle, enter the 1s orbital, and must make do with 2s instead, in spite of the much higher energy entailed. One could in principle continue feeding electrons into Figure 5.6, filling orbitals in order of ascending energy, to build an electronic model of any species of atom, but before we can do so accurately we must recognize two features of multi-electron atoms which have been neglected in Figure 5.6.

In the first place, Figure 5.6 has to be modified a little to allow for electrostatic repulsion between electrons, an effect we did not have to consider in the hydrogen atom. This mutual repulsion in multi-electron atoms, when incorporated into the Schrödinger equation, leads to solutions with a slightly modified energy structure, as shown in Figure 5.7. Orbitals that are not spatially equivalent no longer share the same energy level, but have energies which depend on *l* as well as *n*. Thus although all of the 3d orbitals still have a common energy level (at least in an isolated atom), their energy now exceeds that of the three 3p orbitals, which in turn is greater than the 3s energy. Figure 5.7 provides a more general framework for comparing the electronic structure of different elements. Note that the energy axis in Figure 5.7 is non-linear (see caption).

The second point to be remembered in discussing multi-electron atoms is the increased nuclear charge (owing to the greater number of protons, *Z*, in the nucleus) which causes each electron to experience a stronger electrostatic attraction towards the nucleus. This changes the orbital picture quantitatively but not qualitatively. The shapes of the various orbitals stay the same, but with increasing nuclear charge they all diminish in size as electron density is confined ever more closely to the immediate vicinity of the nucleus. The overall form of Figure 5.7 changes little from element to element, but the negative energy associated with a particular orbital becomes progressively greater (the level becomes 'deeper' in energy space) with increasing nuclear charge, as the energy scales on the left of Figure 5.7 illustrate. In energy terms the two 1s electrons of the uranium atom are held 10 000 times more

Figure 5.7 A generalized diagram of the energy-level structure in more complex atoms. Note that the vertical axis is non-linear (unlike Figure 5.6), as shown by the scales on the left. It approximates to an inverted logarithmic scale.

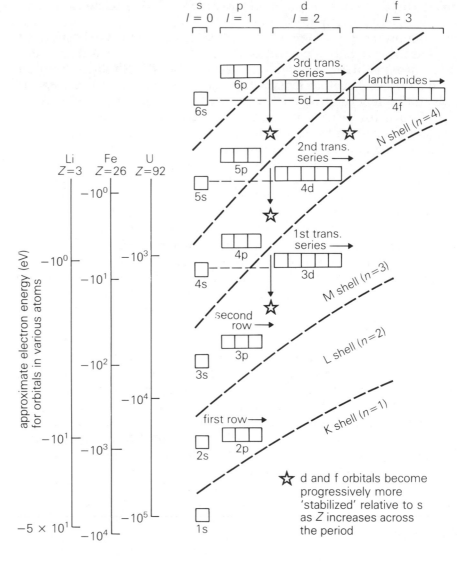

tightly ($\sim 10^5$ electron volts) than is the 1s electron of a lithium atom (55 electron volts).

5.6.2 Electronic configurations

The **electronic** configuration of an atom is a symbolic code describing the location of its electrons in the various orbitals, something one can readily work out from Figure 5.7. The element boron, for instance, has an atomic number (Z) of 5, so that its nucleus contains five protons and has five positive units of charge (5+). Accordingly the boron atom has to accommodate five electrons, and their distribution can be written:

Box 5.4

Chapter 5 – the absolute essentials

Of all the chapters in this book, understanding Chapter 5 requires the most physical insight. The reader is strongly encouraged to persevere until the end of the chapter because, armed with the wave-mechanical view of the atom, you will be best equipped to cope with succeeding chapters of the book. Nevertheless, for the reader to whom the concept of electron 'waveforms' is unacceptably remote from everyday experience, the main conclusions of Chapter 5 are presented below in non-wave terms.

The essential properties of any atom are determined by the number of **protons** (positively charged nuclear particles) in the nucleus. This is known as the **atomic number** Z of the atom.

$Z = 3$, lithium

The value of Z identifies the chemical element to which the atom belongs (Appendix D). The other particles shown in the figure are neutrons (uncharged nuclear particles) that determine the isotopes of the element.

The positive charge ($Z+$) is neutralized by a cloud of Z electrons (negatively charged particles) surrounding the nucleus, held in position by its electrostatic attraction.

The atom forms chemical bonds with other atoms by 'trading' electrons with them. The Z electrons in an atom may be divided into:

(a) **valence electrons**, constituting the 'liquid assets' with which the atom can trade to form bonds with other atoms, and

(b) **core electrons**, the 'capital reserve' of electrons too tightly bound to the nucleus to be involved in bonding.

The number of valence electrons, among other factors, determines the number of bonds that an atom can form (its '**valency**').

Each electron in an atom resides in a specific region of space close to the nucleus called an **orbital**, each orbital accommodating up to two electrons. The various orbitals differ in their symmetry around the nucleus, and this determines the geometry of directional bonds such as those responsible for the structure of diamond.

The size and shape of each orbital is indicated by the values of **quantum numbers** n and l (Table 5.3).

The distribution of electrons between orbitals, and the division into 'core' and 'valence' categories, is determined by the **energy-level diagram** shown in Figure 5.7. Each box represents an orbital accepting up to two electrons. To predict the chemical properties of an atom, Z electrons are 'fed in' from the bottom (lowest energy, most stable) upward. Valence electrons are those residing in the highest occupied energy levels (least strongly bound electrons, most easily removed).

3s 3p

$$\text{boron:} \quad 1s^2 2s^2 2p^1$$

The requirement to minimize total electron energy is satisfied by putting two electrons into the 1s orbital (this is what the code '$1s^2$' means), a further two into 2s (hence '$2s^2$'), and the one remaining electron into one of the three 2p orbitals ('$2p^1$'). We do not need to specify – indeed we have no way of telling – which of the three 2p orbitals receives this single electron.

Another example is the element sodium ($Z = 11$). Its electronic configuration is

$$\text{sodium:} \quad 1s^2 2s^2 2p^6 3s^1$$

In the sodium atom, the three 2p orbitals have accepted their joint quota of six electrons, but one more electron still remains to be accommodated. This goes in the next lowest energy level, 3s. Here however it has a conspicuously higher energy than the other electrons in the atom. Because it extends much further from the nucleus in the 3s orbital, and requires much less energy to remove it from the atom altogether, this solo electron dominates the chemical behaviour of sodium. It is called the **valence electron**, because it can be used in transactions with other atoms, rather like a current account at the bank.

The electrons occupying the 1s, 2s and 2p orbitals of the sodium atom, on the other hand, are much more tightly held and they never participate in sodium's chemical reactions. They comprise what is called the **electron core** of the atom, and resemble personal wealth tied up in stocks and shares, too immobile to be used in day-to-day transactions.

Because of the way orbitals are grouped in energy space by their value of n (Figure 5.7), it is sometimes useful to speak of electron **shells**. Electrons in the 1s state comprise the **K shell**, those in 2s and 2p comprise the **L shell**, and so on, as shown in Figure 5.7. One can likewise speak of valence electrons collectively as the **valence shell**. One must however be careful in attributing spatial significance to the term. It is useful in suggesting that electrons in the M shell project further from the nucleus than those in the L shell, but any impression that electrons are sharply segregated into hollow shells, as some elementary textbooks suggest, is of course false.

5.7 Review

The important conclusions of this chapter may be summarized in the following terms:

(a) The behaviour of the electron, when trapped in an atom, is dominated by its wave-like nature. The electron distributes itself in space about the nucleus rather like a stationary wave on a fixed string.

(b) Each electron in an atom can adopt one of a variety of 'waveforms' having different spatial configurations of electron density. These are called orbitals. Each corresponds to a solution of the Schrödinger wave equation. Its size and shape are defined by the values of various quantum numbers, in a manner reminiscent of the harmonics of a vibrating string.

(c) Electron energy in the atom is quantized, like the frequency of a guitar string (and for the same reason). Each orbital has its own discrete energy level, and collectively they give rise to an electron energy structure resembling an irregular set of pigeonholes (Figure 5.7), which in qualitative terms is common to all types of atom.

(d) Two electrons can share the same orbital provided that they have opposed 'spins'. Each electron normally occupies the lowest-energy orbital in which space is available. Energy considerations thus dictate the geometrical distribution of electron density about the nucleus.

As the following chapters will show, these principles, together with the energy level scheme shown in Figure 5.7, form the foundation upon which modern inorganic chemistry is built.

5.8   Further reading

Fyfe, W. S. (1964) *Geochemistry of solids. An introduction*, McGraw Hill, New York. (Chapter 2.)
Shriver, D. F., Atkins, P. W. and Langford, C. H. (1994) *Inorganic chemistry*, 2nd edn, Oxford University Press, Oxford.

5.9   Exercises

1 Work out which electron orbitals have the following quantum numbers:

n	l
2	1
3	0
4	3
5	2

2 Determine the electron configurations of the chemical elements having the atomic numbers 6, 11, 13, 17, 18, 26.

6
The Periodic Table

A chemical element is identified by its atomic number Z, which defines both the number of protons in the nucleus (and thus the nuclear charge), and the number of electrons in the neutral atom (Box 6.1). In the present chapter we consider how the atomic number, in conjunction with the electron energy structure developed in Chapter 5 (Figure 5.7), determines the chemical properties of the element concerned. The structure of the energy-level diagram leads to a **periodic repetition** of chemical properties, which is conveniently summarized by tabulating the elements on a grid known as the Periodic Table (p. 299).

Though the architecture of the Periodic Table can be seen as a consequence of wave-mechanical theory, it was originally worked out from chemical observation. It was first published in its modern form by Dimitri Mendeleev in 1869, nearly sixty years before the work of Schrödinger.

6.1 Ionization energy

The bonds formed by an atom involve the transfer or sharing of electrons. It therefore makes sense to illustrate the periodicity of chemical properties by looking at a parameter that expresses how easy or difficult it is to remove an electron from an atom. The **ionization energy** of an element is the energy input (expressed in $J\ mol^{-1}$) required to detach the loosest electron from atoms of that element (in its ground state). It is the energy difference between the 'free electron at rest' state (the zero on the scale of electron energy levels) and the highest occupied energy level in the atom concerned. What this means in the simplest case, the hydrogen atom, is shown in Figure 5.6. A low ionization energy denotes an easily removed electron, a high value a strongly held one.

We can picture how ionization energy will vary with atomic number

Box 6.1

Chemical symbols

A few of the one- or two-letter codes for the chemical elements will be familiar to most readers. Most are abbreviations of the English element names, but a few refer to Latin names such as Na (natrium) for sodium and Ag (argentum) for silver, or to alternative continental names like 'wolfram' for tungsten (W). A list of chemical symbols is given at the end of the book (Appendix D).

Using subscripts and superscripts, chemical symbols can be augmented to specify every detail about a particular atom:

The mass number $A = (Z + N)$ specifies one isotope of the element.	The charge on the ionized atom. (Alternatively the oxidation state can be indicated in capital Roman numerals.)

$$\mathrm{^{57}_{26}Fe^{3+}_{2}}$$

The atomic number (Z).	The number of atoms in the molecule under consideration.

It is usually sufficient to specify only one or two of these numbers. They should always be written in the positions shown, to avoid ambiguity. For example, to specify a particular isotope (Chapter 10) one writes ^{40}Ar. (Older literature uses the obsolete notation Ar^{40}.)

The atomic number Z can generally be omitted, its value being implied by the chemical symbol itself (but see Figure 6.2).

---

by considering the highest occupied energy level in each type of atom (Figure 5.7). In lithium (Li; $Z = 3$, $1s^22s^1$) and beryllium (Be; $Z = 4$, $1s^22s^2$) it is the 2s level; in boron (B; $Z = 5$, $1s^22s^22p^1$) it is the 2p level; and so on. If we were to disregard the increasing nuclear charge, we would predict that the energy needed to strip an electron from this 'outermost' level would vary with atomic number as shown in Figure 6.1a. One would expect a general decline in ionization energy with increasing Z, punctuated by sudden drops marking the large energy gaps between one 'shell' and the next one up (Figure 5.7); the downward series of steps in Figure 6.1a thus reflects the occupation of progressively higher energy levels in Figure 5.7. There is no suggestion of periodicity.

Because nuclear charge, and therefore the strength of the nuclear field, increase with atomic number, we find that each of the level steps expected in Figure 6.1a is actually a ramp (Figure 6.1b), whose rising

Figure 6.1 a. A plot of ionization energy against atomic number predicted without regard to the effect of increasing nuclear charge. b. The variation of measured ionization energy with atomic number Z among the first 20 elements. (The whole Z-range is shown in the inset.) The rising profile between each abrupt drop reflects the increasing nuclear charge.

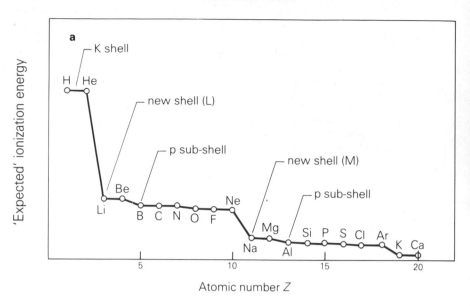

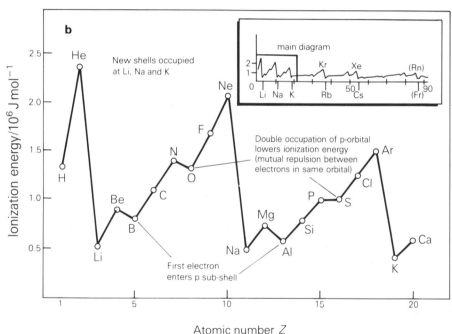

profile reflects the increase in nuclear attraction experienced by each electron in the atom in passing from one atomic number to the next. The ionization energy of helium is (nearly) twice that of hydrogen, for example, because its doubly charged nucleus attracts each electron twice as strongly as the singly charged hydrogen nucleus. The ramps are separated by the sudden drops noted in Figure 6.1a, producing a markedly periodic variation of ionization energy.

At the top of each ramp is an element which hangs on tenaciously to all of its electrons. These elements are the **noble gases**, helium (He), neon (Ne) and argon (Ar). (Two others, krypton (Kr) and xenon (Xe), lie beyond the Z-range of the diagram – Figure 6.1b inset). Their electronic structures are characterized by completely filled shells, in which all electrons are held so firmly that the exchange of electrons involved in chemical bonding is ruled out. Noble gases therefore exhibit no significant chemical reactivity. Indeed the electronic structure of the noble gases is so stable that other elements seek to emulate it by losing electrons, or by acquiring additional electrons. Instead of forming **diatomic** molecules like O_2, N_2 and Cl_2, the noble gases are **monatomic**.

Immediately to the right of each noble gas in Figure 6.1b lies an element with a conspicuously low ionization energy. Lithium (Li), sodium (Na) and potassium (K) are **alkali metals**, whose electronic structures consist of the filled shells of the preceding noble gas, plus one further electron which has to occupy the next shell at a significantly higher energy (Figure 5.7). It projects further from the nucleus than the core electrons, and is **screened** by them from the full attraction of the nuclear charge, making it easier to remove or involve in bonding. The chemistry of the alkali metals is dominated by this single **valence electron**. Removing a second electron would be much more difficult, because to do so would mean breaking into the stable noble-gas core (Chapter 5). Because of their low ionization energies, the alkali metals readily form singly charged M^+ cations.

Beryllium (Be; $1s^2 2s^2$), magnesium (Mg; $1s^2 2s^2 2p^6 3s^2$) and calcium (Ca; $1s^2 2s^2 p^6 3s^2 3p^6 4s^2$) each have two electrons in the **valence shell**, both of which are fairly easy to remove (though not as easy as the single electron in the valence shell of an alkali metal). These **alkaline earth metals** utilize both of these electrons in their bonding and, with the exception of Be (Chapter 8), readily form the doubly charged M^{2+} cation.

Boron (B) and aluminium (Al), to the right of Be and Mg in Figure 6.1b, each have three valence electrons to utilize in their chemical reactions; and carbon (C) and silicon (Si) each have four. The increasing nuclear attraction makes these electrons progressively harder to remove, however, and the tendency among these elements to form cations is significant only for Al; the chemistry of B, C and Si is dominated by bonds in which electrons are shared (Chapter 7).

The periodic pattern becomes more complicated at Z values above 20 (inset in Figure 6.1b), owing to the presence of electrons in d-orbitals. The overall periodicity nevertheless persists, the minimum ionization energies belonging to the alkali metals rubidium (Rb) and caesium (Cs), and the maximum values coinciding with the noble gases krypton (Kr) and xenon (Xe).

6.2 The Periodic Table

Having arranged the elements in order of atomic number, they can be divided into several **periods**, each beginning with an alkali metal – except the first, which begins with hydrogen – and concluding with a noble gas. Writing successive periods underneath each other on a page gives a coherent layout of the elements that highlights their common chemical properties. This is shown for the first three periods (including hydrogen and helium) in Figure 6.2. Notice that each of the groups of similar elements that emerged from Figure 6.1b now forms its own column. The first contains the alkali metals Li, Na and K, to which by tradition we add the element hydrogen, because it resembles the alkali metals in having just one electron in the valence shell. In column 2 we find the alkaline earth metals, Be, Mg and Ca; in column 3, B and Al, and so on.

Numbering the columns in Figure 6.2 from left to right divides the elements into eight groups, identified by Roman numerals, according to the number of electrons in the valence shell. Numbering the periods indicates the n-value (the principal quantum number) of the current valence shell: thus for Al in period 3 the valence electrons are in orbitals that have $n = 3$ (the M shell). The noble gases at the end of each period mark the **closure** (filling up) of the p-orbitals of that shell. It is therefore logical to place helium at the head of this eighth column rather than column 2, as its two electrons close the K (1s) shell.

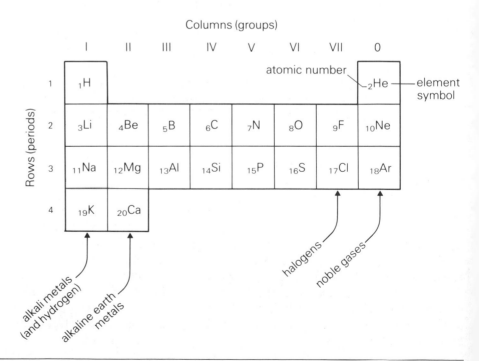

Figure 6.2 A condensed Periodic Table for the first 20 elements. Chemical symbolism is discussed in Box 6.1.

This is the rationale of the **Periodic Table**, which is shown in its complete form on p. 299. Notice that it has been split between Groups II and III in order to accommodate elements like scandium (Sc), titanium (Ti) and iron (Fe). This is where electrons begin to occupy the lowest d-orbitals (3d). The ten elements from Sc to Zn (zinc) comprise the First Transition Series. They include many industrially important metals.

The next six elements, from gallium (Ga) to the noble gas krypton (Kr), correspond to the filling of the 4p orbitals, the next sub-shell in order of energy (Figure 5.7).

Subsequent periods follow the same pattern, except for the added complication of the availability of 4f and 5f orbitals in very heavy atoms. The entry of electrons into 4f orbitals generates a series of geochemically important trace elements called the **lanthanides** (cerium, Ce, to lutetium, Lu), though in geological literature they are more commonly known – together with the preceding element lanthanum, La – as the **rare earth elements** (REE – see p. 229).

In geological terms, the most important important heavy elements of all are thorium (Th) and uranium (U). Owing to their radioactivity, they make a major contribution to heat generation in the Earth, and are important in geochronology. They belong to a similar series of elements arising from the filling of 5f orbitals, known as the **actinides** (Chapter 9). None have stable nuclei, only Th and U having sufficiently long half-lives to occur in nature today (Chapter 10).

6.3 Electronegativity

Ionization energy provides a useful indication of the periodicity of element behaviour, but its chemical applications are limited to elements that lose electrons to form cations. **Electronegativity** is a more versatile concept, summarizing the chemistry of all kinds of elements and the bonds they form. Electronegativity is a number that indicates the capacity of an atom in a molecule or crystal to attract extra electrons. In the alkali metals, this capacity is hardly developed at all; elements that tend to give electrons away rather than attracting additional ones are called **electropositive** elements. They appear on the left-hand side of the Periodic Table. Their electronegativity values are low, beginning at 0.8 (for the alkali metals K, Rb and Cs).

The most **electronegative** elements are those having nearly complete valence shells, on the right-hand side of the Periodic Table. The nuclear charge effect draws the valence orbitals closer to the nucleus, offering an incoming electron a state of low energy, compared for example with the valence shell of an alkali metal where the attraction of the nucleus is only weakly felt. Electronegative atoms have the power to attract and retain additional electrons, in spite of the net negative charge which the atom thereby acquires (making it an anion). The explanation for this

133

Figure 6.3 Variations of electronegativity (height of each block) across the Periodic Table (stable elements only). Groups and periods are shown in Roman and Arabic numerals respectively. (Data source: P. Henderson (1982) *Inorganic Chemistry*, Pergamon, Oxford.) The gap between molybdenum (Mo) and ruthenium (Ru) belongs to the radioactive element technetium (Tc–see Exercise 4).

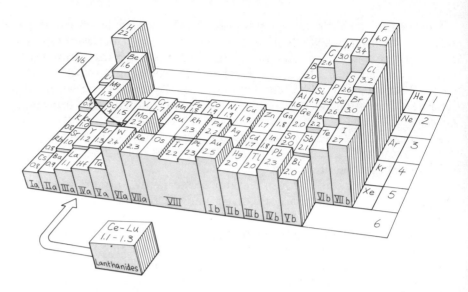

phenomenon is partly wave-mechanical, to do with the special stability of the noble-gas configuration which such an atom can achieve with the aid of borrowed electrons. The most electronegative element is fluorine (F, electronegativity 4.0).

Figure 6.3 shows that electronegativity varies in a fairly regular manner in the Periodic Table. It increases strongly from left to right, and more gently from bottom to top (although the latter trend is reversed in the central area). As a rule, metals have electronegativities less than 2.0, whereas non-metals have values greater than 2.5.

6.4 Valency

The number of bonds that an atom can form as part of a compound is expressed by the **valency** of the element. In chemical reactions, elements adjust their electron populations to achieve a noble-gas configuration. Because the single valence electron of alkali metals like Na ($1s^2 2s^2 2p^6 3s^1$) allows them to form only one bond (Chapter 7), they are said to be monovalent (valency = 1). Magnesium ($1s^2 2s^2 2p^6 3s^2$) has a valency of two (is 'divalent'), because two of the electrons in the neutral Mg atom are in the valence shell, and can be used to establish bonds. For strongly electropositive elements like these (Figure 6.3), valency is equal to the number of electrons in the valence shell, and can therefore be determined from the column in the Periodic Table in which the element occurs. The valencies of the elements B and Al (trivalent), C and Si (quadrivalent) and P (pentavalent) also conform to their position in the Periodic Table.

The electronegative elements like oxygen and chlorine require a complementary definition of valency. These are elements with nearly complete valence shells. Because they can achieve a noble-gas electron structure by accepting extra electrons from other atoms (or sharing them), it is the number of vacancies in the valence shell that determines the number of bonds that can be formed. Oxygen, with six valence electrons and two vacancies, can establish two bonds with other atoms and is divalent. Chlorine (seven valence electrons, one vacancy) needs only one electron to complete the valence shell and is therefore monovalent.

Many of the elements in the central parts of the Periodic Table behave in a more complicated manner. In Chapter 4 we discussed several elements that can adopt one of several oxidation states depending on the oxidizing or reducing properties of the environment. Each of these oxidation states represents a separate valency of the element concerned. Multiple valency is most characteristic of the transition elements (Figure 9.7). The best-known example is iron, which in addition to the metallic state (valency 0) can exist in the geological world as ferrous (divalent) or ferric (trivalent) compounds. It often happens, as in the case of iron, that none of the oxidation states corresponds to the valency suggested by the element's position in the Periodic Table. Iron occurs in Group VIII (Box 6.2), but exhibits valencies of 2 and 3. It is clear that such elements do not utilize all of the electrons that nominally belong to the valence shell in forming bonds, for reasons which will be examined in Chapter 9. The same is true of many heavier elements in the p-block (Box 6.2). Tin, for example, occurs in the geological environments as Sn (II) and Sn (IV) compounds.

Box 6.2

Sub-groups and blocks

The condensed form of the Periodic Table for the first twenty elements (Figure 6.2) is easy to divide into eight columns, but the introduction of the transition series – an additional ten columns – makes necessary a revision of these column headings. In order that elements like B, C and Al retain group numbers reflecting their valency, columns are divided into 'a' and 'b' sub-groups as shown on p. 299, leaving three columns in the middle of the transition series (headed by Fe, Co and Ni) which are lumped together as Group VIII. Formal chemical similarities exist between corresponding a and b sub-groups (e.g. valency, as in sub-groups IVa and IVb), but they are generally outweighed by the differences in electronegativity (Figure 6.3).

It is useful to divide the Periodic Table into blocks reflecting the kind of sub-shell (s, p, d or f) that is currently incomplete or just filled. As shown on p. 299, the transition series comprise the d-block and the lanthanides and actinides the f-block. It is important to recognize the direct relationship of these blocks, and of the structure of the Periodic Table as a whole, to the energy-level diagram shown in Figure 5.7.

6.5 Atomic spectra

An atom is said to be in its **ground state** when all of its electrons occupy the lowest energy levels allowed to them by the Pauli Principle (Chapter 5). This lowest-energy configuration (Figure 6.4) is the one normally encountered at room temperature. But atoms can absorb energy from their surroundings, for example when they are heated or exposed to energetic radiation, causing one or more electrons to jump from a stable, low energy level into one of the vacant orbitals at higher energy, or perhaps even to be ejected from the atom altogether. This unstable **excited state** of the atom, with a vacancy in a low energy level, soon reverts to the stable ground state by filling the vacancy with an electron from a higher level. The electron making the downward transition must dispose of an amount of energy ΔE (Figure 6.4) equal to the difference between its initial and final energy levels, and this energy output takes the form of electromagnetic radiation (Box 6.3). Excited atoms therefore emit a series of sharply defined wavelength

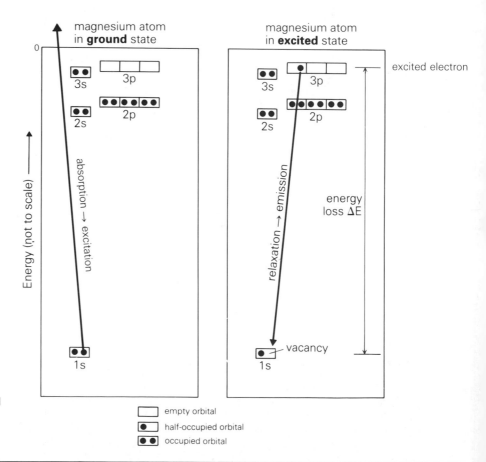

Figure 6.4 Ground states and excited states. The transition shown emits an MgKβ quantum as an X-ray photon.

Box 6.3

Light and other forms of electromagnetic radiation

The light that we see, like other forms of electromagnetic radiation, is an electromagnetic disturbance which propagates energy through space, rather like the radiating ripples on the surface of a pond disturbed by a stone. The source of light excites simultaneous 'ripples' in electric and magnetic field strength, which spread out from the source at the speed of light.

The essential characteristics of any electromagnetic wave are the **frequency** of vibration of the electromagnetic field (v) in hertz (Hz = s^{-1}) and its **wavelength** (λ) in metres (Box 5.2). These are complementary properties related through the equation:

$$\lambda v = c$$

where c is the speed of light in m s^{-1} (c = 2.997 $\times$ 10^8 m s^{-1} in vacuum). The wavelength is the parameter normally used to characterize the quality of visible light that we call colour. Frequency, however, is the more fundamental property: unlike wavelength and c, it is independent of the refractive index of the medium through which the light is passing.

Light energy is **quantized**: a light beam, though apparently a continuous stream of waves, consists of minute packets or 'quanta' of wave energy called **photons**, resembling the wave pulses associated with the electron (Figure 5.1). Planck showed at the turn of the century that each photon has a kinetic energy E_q related to the frequency of the light of which it forms a part:

$$E_q = hv$$

where h is called Planck's constant and has the value 6.626 $\times$ 10^{-34} J s.

When an electron falls from a high energy level in an atom to a lower one, it emits a quantum of energy in the form of an electromagnetic photon, whose energy is equal to the **energy difference** ΔE between the electron's initial and final states. It follows that the light emitted by atoms undergoing this transition has a frequency given by:

$$v = E_q/h = \Delta E/h$$

The corresponding wavelength is $\lambda = \dfrac{hc}{\Delta E}$

Since the energy levels (and ΔEs) in an atom depend on the nuclear charge Z, the wavelengths of atomic spectra vary predictably from one element to the next, and can be used (when separated by a spectrometer into constituent wavelengths) to identify the elements present in a complex sample without separating the elements chemically. The **intensity** of each wavelength 'peak' in the spectrum provides a measure of the concentration in the sample of the element to which it relates (Box 6.4, p. 140).

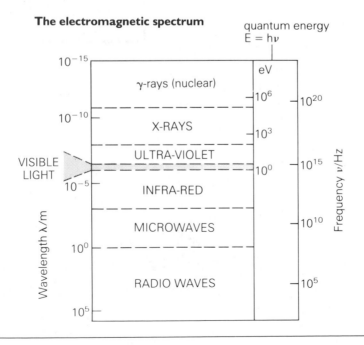

The electromagnetic spectrum

peaks (Figure 6.5) that provide detailed information about their electronic energy structure: these wavelengths constitute the **electromagnetic spectrum** of the element(s) concerned. Because the energy levels of an atom, and therefore the wavelengths it emits, are Z-dependent, the spectrum of one element is readily distinguishable from that of another (Box 6.3). The success in explaining why an atomic spectrum consists of a series of sharp lines rather than a continuum is one of the triumphs of wave mechanics.

Can an electron leap from any energy level to any other level within an atom? Analysis of the peaks present in an atomic spectrum indicates that the answer must be 'no'. Certain transitions are 'forbidden' because they would violate basic physical principles such as the conservation of

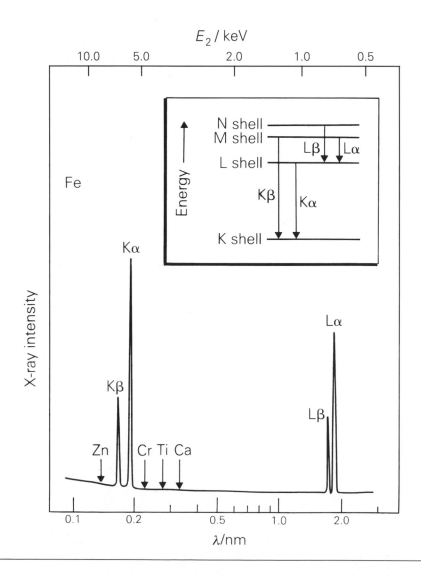

Figure 6.5 The X-ray spectrum of iron as shown by an X-ray spectrometer. The inset shows the electron transitions involved. The arrows in the main diagram show how Kα wavelength shifts from element to element. The width of the peaks has been exaggerated.

angular momentum. Wave mechanics recognizes such restrictions in the form of a number 'selection rules'. For example, a radiative transition in an atom must satisfy the two conditions:

$$\Delta l = \pm 1$$

$$\Delta n \neq 0$$

Thus element spectra do not include lines that correspond to transitions between 3s and 2s states (for which $\Delta l = 0$), between 3d and 2s (for which $\Delta l = 2$), or between 3p and 3s (for which $\Delta n = 0$).

6.5.1   X-ray spectra

X-rays are electromagnetic waves of very short wavelength (about 10^{-8} to 10^{-11} m) and high frequency (Box 6.3). They are the most energetic form of radiation that can be generated in the electron shells of atoms. (γ-rays have higher energies, but are produced in the nucleus.) The high energies indicate that X-rays arise from electron transitions involving the deepest, most tightly bound energy levels in the atom, in particular the K and L shells. The energy level structure in these shells is simple (Figure 5.7) owing to the restrictions that apply to the value of quantum number l when n is small. X-ray spectra therefore consist of relatively few lines, a fact which makes them convenient for the analysis of complex, multi-element samples like rocks and minerals. In common with all atomic spectra, the wavelength of each X-ray peak depends on the atomic number of the element emitting it (Figure 6.5).

The production of X-rays involves creating a vacancy in the K or L shell by ejecting an electron completely out of the atom (Figure 6.4). Two methods can be used to excite atoms into generating X-ray spectra:

(a) A very narrow beam of high-energy electrons can be focused on a small area on the surface of the sample (usually a crystal on the surface of a polished thin section). The energy of the electrons excites the atoms with which they collide in the tiny volume of the sample directly under the area of impact, from which the X-ray spectra of elements present in the sample are emitted. This is the principle of the **electron microprobe** (Box 6.4), widely used for the chemical analysis of minerals in geological thin sections.

(b) A powerful beam of X-rays (from an X-ray tube), when directed at a sample in powdered or fused form, will prompt X-ray emission in the sample by **fluorescence**. The photon energy of the incoming ('primary') X-rays, E_q, must be sufficient to eject the relevant electrons from the sample atoms. The sample responds by emitting 'secondary' or 'fluorescent' X-ray spectra characteristic of the elements present in the sample. X-ray fluorescence

Box 6.4

The electron microprobe

The electron microprobe is a scanning electron microscope in which the electron beam is used to produce X-rays instead of to produce images. The technique has revolutionized petrology since it was introduced in the 1950s.

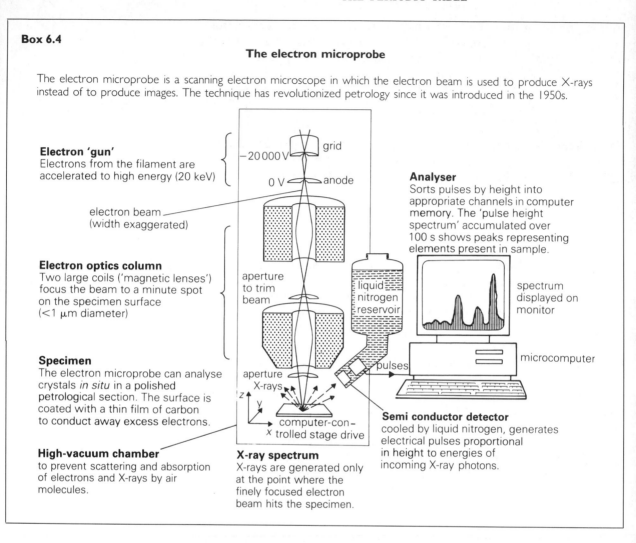

Electron 'gun'
Electrons from the filament are accelerated to high energy (20 keV)

−20 000 V   grid

0 V   anode

electron beam
(width exaggerated)

Electron optics column
Two large coils ('magnetic lenses') focus the beam to a minute spot on the specimen surface (<1 μm diameter)

aperture
to trim
beam

liquid nitrogen reservoir

Analyser
Sorts pulses by height into appropriate channels in computer memory. The 'pulse height spectrum' accumulated over 100 s shows peaks representing elements present in sample.

spectrum displayed on monitor

microcomputer

Specimen
The electron microprobe can analyse crystals *in situ* in a polished petrological section. The surface is coated with a thin film of carbon to conduct away excess electrons.

aperture
X-rays

computer-con−
trolled stage drive

pulses

Semi conductor detector
cooled by liquid nitrogen, generates electrical pulses proportional in height to energies of incoming X-ray photons.

High-vacuum chamber
to prevent scattering and absorption of electrons and X-rays by air molecules.

X-ray spectrum
X-rays are generated only at the point where the finely focused electron beam hits the specimen.

spectrometry is an important rapid method of whole-rock analysis, applicable to major elements and many trace elements.

Z-DEPENDENCE OF X-RAY SPECTRA: MOSELEY'S LAW

One advantage of using X-ray spectra for rock and mineral analysis is that element wavelengths depend in a very simple way on the atomic number Z. Increasing the nuclear charge stretches the energy level structure downward in energy space, expanding the energy differences ΔE between different levels. This is easily seen from the energy scales in Figure 5.7. It follows that the photon energy for a given transition increases with atomic number, while the corresponding wavelength decreases (Figure 6.5). This relationship is expressed in a simple equation established empirically by the British physicist H. G. J. Moseley in 1914 and known as **Moseley's Law:**

$$\frac{1}{\lambda} = k(Z - \sigma)^2 \qquad (6.1a)$$

Alternatively this may be written in terms of photon energy E:

$$\frac{E}{hc} = k(Z - \sigma)^2 \qquad (6.1b)$$

The **screening constant**, σ, represents the degree to which inter-electron repulsion diminishes the attraction of the positive nuclear charge. $Z - \sigma$ can be regarded as the 'effective nuclear charge' felt by an individual electron. Moseley, who was the first to introduce the concept of atomic number, used this equation to catalogue the chemical elements known in 1914, and to demonstrate from gaps in the sequence of X-ray wavelengths that others (with atomic numbers of 43, 61, 72, 75, 85 and 87) still remained to be discovered.

For plotting the Moseley equation (6.1) in a graph, it is convenient to express it in terms of the square roots of each side:

$$\left(\frac{1}{\lambda}\right)^{0.5} = k^{0.5}(Z - \sigma) \qquad (6.2a)$$

or

$$(E_q)^{0.5} = (hck)^{0.5}(Z - \sigma) \qquad (6.2b)$$

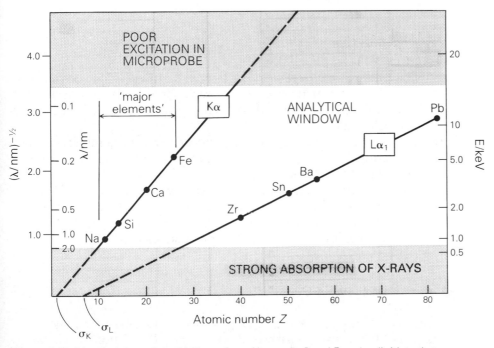

Figure 6.6 Moseley's Law plotted in linear form (Appendix B and Exercise 4). Note that the vertical axis is graduated in $\lambda^{-\frac{1}{2}}$ (linear scale) and λ. The unstippled area shows the range of wavelengths and elements attainable in routine microprobe analysis of minerals and rocks.

Plotting $(1/\lambda)^{0.5}$ or $E_q^{0.5}$ against Z will produce a straight line with a gradient of $k^{0.5}$ or $(hck)^{0.5}$ respectively (Figure 6.6), whereas plotting $1/\lambda$ or E_q directly versus Z would have given a less useful curved line (Appendix B).

The same equation can be used for predicting other X-ray lines such as Kβ and Lα for particular element, but the constants k and σ will have different values (Figure 6.6).

The analysis of X-ray spectra can be carried out in practice by one of two methods. A crystal of known atomic spacing can be used as an X-ray diffraction grating (Box 5.3) to disperse the various wavelength components of the spectrum into a series of peaks (wavelength-dispersive or 'WD' analysis) whose intensities can be recorded by a relatively simple X-ray detector driven mechanically to the appropriate angles. Alternatively the incoming X-ray beam can be passed directly into a semiconductor detector that can separate the spectral components according to photon energy (energy-dispersive or 'ED' analysis; Box 6.4). How a semiconductor detector works is explained in Chapter 7.

It can be seen from Figure 6.6 that the long-wavelength X-ray spectra from elements of atomic number less than 10 are strongly absorbed, and this limits the effectiveness of X-ray methods when analysing elements having atomic numbers lower than 10, though modern spectrometer design has greatly improved light-element performance in recent years.

6.6   Review

(a)   The energy level structure in atoms (Figure 5.7) in conjunction with the effect of increasing nuclear charge leads to a periodicity of chemical properties when elements are examined in atomic-number order.

(b)   The Periodic Table provides a concise means of summarizing and predicting the variation of chemical properties (such as electronegativity and valency) among chemical elements.

(c)   Wavelengths of elemental spectra vary systematically with atomic number, providing a powerful means of analysing the elements present in a geological sample. X-ray spectra are particularly useful for the chemical analysis of minerals and rocks.

6.7   Further reading

Gill, R. (ed.) (1996) *Modern analytical geochemistry*, Longman, Harlow.

Puddephat, R. J. and Monaghan P. K. (1986) *The Periodic Table of the elements*, 2nd edn, Oxford University Press, Oxford.

Shriver, D. F., Atkins, P. W. and Langford, C. H. (1994) *Inorganic Chemistry*, 2nd edn, Oxford University Press, Oxford.

6.8 Exercises

1 Identify the elements having the atomic numbers listed below. Work out their electronic configurations, distinguishing between core and valence electrons. Establish the block and group to which each element belongs, and work out the valency.

$$Z = 3, 5, 8, 9, 14$$

2 Work out the electronic configurations of the following atoms, representing the electron core by the symbol of the preceding inert gas. To which blocks do they belong?

$$\text{Ti Ni As U}$$

3 Determine the (minimum) values of x and y in the following formulae, consistent with the valencies of the elements concerned:

$$\text{Na}_x\text{O}_y \ \text{Si}_x\text{O}_y \ \text{Si}_x\text{F}_y \ \text{Mg}_x\text{Cl}_y \ \text{Sc}_x\text{O}_y \ \text{P}_x\text{O}_y \ \text{B}_x\text{N}_y$$

4 (a) Use the $K\alpha$ wavelength data below to plot a graph to verify Moseley's Law for the elements Y to Ag (see Appendix B and Figure 6.6). Estimate the values of k and σ.

Element	Z	Wavelength ($\text{Å} = 10^{-10}\,\text{m}$)
Y	39	0.830
Zr	40	0.787
Nb	41	0.748
Mo	42	0.711
Ru	44	0.644
Rh	45	0.615
Pd	46	0.587
Ag	47	0.561

(b) The radioactive element technetium (Tc, $Z = 43$; named from the Greek *technetos* = 'artificial' – Figure 6.3) does not occur naturally on Earth, but can be produced artificially. Predict the wavelength of its $K\alpha$ X-ray line and the corresponding quantum energy (in keV). ($h = 6.626 \times 10^{-34}\,\text{J s} = 4.135 \times 10^{-15}\,\text{eV s}$; $c = 2.997 \times 10^{8}\,\text{m s}^{-1}$ in vacuum.)

7
Chemical bonding and the properties of minerals

Few scientists are expected to deal with as wide a range of materials and properties as the geologist. Consider the contrast between red-hot silicate lava and grey Atlantic sea water, between the engineering properties of crystalline granite and those of soft clay or mud, between the electrical and optical properties of quartz and those of gold. The immense physical diversity of geological materials is derived largely from the differences in the chemical bonding that holds them together.

One can distinguish several different mechanisms of chemical bonding, though the real interaction between two atoms is generally a mixture of more than one bonding type. The extent to which each mechanism contributes to a real bond depends on the difference in electronegativity between the atoms concerned. We begin by examining the type of bond that predominates when the electronegativity contrast is large.

7.1   The ionic model of bonding

The salt sodium chloride, familiar as table salt and as the mineral halite, consists of two elements of notably different electronegativity: 3.2 (Cl) and 0.9 (Na) (difference = 2.3). The low ionization energy of the sodium atom (Chapter 6) indicates a readiness to lose an electron, forming the Na^+ cation. The chlorine atom, on the other hand, readily accepts an extra electron, forming the chloride anion Cl^-. When a sodium atom encounters a chlorine atom, one electron may be drawn from the exposed Na 3s orbital into the vacancy in the Cl 3p orbital. The ions resulting from this transfer, having opposite charges, experience a mutual attraction that we call **ionic bonding**.

Electrostatic forces operate in all directions, and an ion in an ionic compound draws its stability from the attraction of all oppositely charged ions nearby. Ionic bonding does not lead to the formation of discrete molecules like CO_2. Ionic compounds exist as solids or liquids ('condensed' phases), which optimize their stability by packing oppositely charged ions closely together in extended structures, or as ionic solutions in which they are stabilized by being surrounded by polar solvent molecules (Box 4.1). Ionic compounds do not exist as gases.

7.1.1 Ionic crystals: stacking of spheres in three dimensions

One can think of most ions as being spherically symmetric. The internal architecture of crystals like NaCl can be understood in terms of stacking spheres of different sizes and different charges into regular three-dimensional arrays. The potential-energy equation for such an array will be the grand sum of (i) negative terms representing the attractive force acting between all pairs of oppositely charged ions in the structure, and (ii) positive terms representing repulsion between all pairs of similarly charged ions. There are three general rules to observe for achieving maximum stabilization (i.e. minimum potential energy):

(a) Ions must obviously combine in proportions leading to an electrically neutral crystal. A halite crystal contains equal numbers of Na^+ and Cl^- ions, whereas in fluorite there must be twice as many F^- ions as Ca^{2+} ions for the charges to cancel.
(b) The spacing between neighbouring oppositely charged ions should approximate to an equilibrium bond length r_0 for the compound concerned (Box 7.1), so maximizing the attractive forces holding the structure together. Stretching a bond makes it less stable.
(c) Each cation should be surrounded by as many anions as their relative sizes will allow. This achieves the maximum degree of cation–anion attraction. For the same reason, each anion needs to be closely surrounded by as many cations as possible. The number of oppositely charged nearest neighbours surrounding an ion (in three dimensions) is called its **co-ordination number**, an important parameter in crystal chemistry.

These rules indicate that the atomic arrangement found in a crystal of halite or fluorite is determined primarily by the charge and the size of the constituent ions. The charge can be predicted from the ion's valency, but how can its 'size' be established?

IONIC RADIUS
Because ions have fuzzy outlines (Figures 5.3 and 5.4), we face a problem in defining precisely what we mean by the 'radius' of an ion.

Box 7.1

Equilibrium inter-nuclear distance

Two oppositely charged ions (for example, Na^+ and Cl^-) are attracted towards each other by their opposite net charges ($\pm e$) and the attraction increases as they get closer. The potential energy of this system is:

$$E_p = -\frac{e^2}{r}$$

As r, the distance between the two nuclei, gets smaller, the potential energy becomes more negative, indicating greater stability. When the ions get so close to each other that the negative electron clouds begin to intermingle, however, we have to allow for an element of repulsion in the energy equation:

$$E_p = \quad -\frac{e^2}{r} \quad + \quad \frac{be^2}{r^{12}}$$

$$\underset{\text{long-range}}{} \quad \underset{\substack{\text{long-range}\\\text{attraction}}}{\phantom{-\frac{e^2}{r}}} \quad \underset{\substack{\text{short-range}\\\text{repulsion}}}{\phantom{\frac{be^2}{r^{12}}}}$$

where b is a constant. The dashed lines in the diagram show the two terms plotted separately. The r^{-12} term indicates that repulsion is felt over a much shorter inter-nuclear distance than the attractive force, but rises very steeply as r is reduced below a critical value.

The solid curve shows the result of adding the two terms together. The energy minimum marks an equilibrium inter-nuclear distance or 'bond-length' r_0 at which the isolated ion pair is most stable. (When the cation is associated with more than one anion, as in a crystal, repulsion between the anions makes r_0 somewhat larger.) The rapid rise of the curve to the left of this minimum reflects the observation that the ions strongly resist further compression, rather like hard rubber balls.

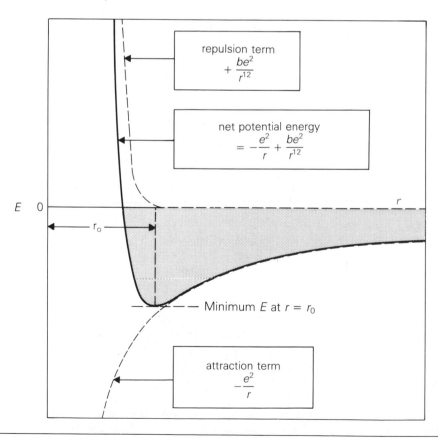

repulsion term
$$+\frac{be^2}{r^{12}}$$

net potential energy
$$= -\frac{e^2}{r} + \frac{be^2}{r^{12}}$$

E 0

r_0

Minimum E at $r = r_0$

attraction term
$$-\frac{e^2}{r}$$

Nevertheless, when two oppositely charged ions come into contact they establish a well-defined equilibrium bond-length (Box 7.1). The bond-length can be regarded as the sum of two hypothetical 'ionic radii', one for each individual ion, as shown in Figure 7.1. It will be clear that the 'radius' of an ion in a crystal represents not the actual size of the isolated ion (whatever that may mean), but an empirical 'radius of approach' governing how close to it another ion can be placed.

X-ray diffraction data (Box 5.3) provide accurate equilibrium bond-lengths for many binary (two-element) ionic salts like NaCl, NaF and CaF_2. The problem in determining the individual ionic radii of Na^+ and Cl^- lies in deciding how much of the Na–Cl bond-length to attribute to each ion, the radius of neither being known initially. An early approach to the problem, devised by the American Nobel Prize-winning chemist Linus Pauling in 1927, was to examine compounds

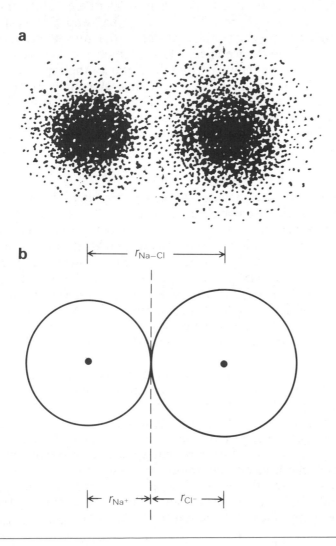

Figure 7.1 Inter-nuclear distance r_{Na-Cl} and ionic radii r_{Na^+} and r_{Cl^-} for the ion pair Na^+Cl^-. a. Electron density distributions of the two ions. b. Corresponding hard-sphere approximation.

like NaF in which the cation and anion happen to have the same number of electrons (10 in this case). Pauling suggested a relationship between ionic radius and nuclear charge in such ion pairs which, together with the measured Na–F bond-length, allowed the separate radii to be estimated (Fyfe, 1964, Chapter 4).

Modern ionic radii are based on more elaborate calculations, though chemists are still not universally agreed on the best values to use: recent estimates of the O^{2-} radius, for example, vary between 1.27 and 1.40 Å (Henderson, 1982, Chapter 6). Mineralogists and geochemists nevertheless find ionic radii extremely useful in explaining the chemical make-up of crystalline materials. The ionic radii of some geologically important elements are illustrated in Box 7.2.

One can see from Box 7.2 that cation radii vary considerably, from 0.34 Å (Si^{4+}) to about 1.7 Å (Rb^+). Note the marked decrease in cation radius in proceeding from left to right in the Periodic Table, in response to increasing nuclear charge. Na^+, Mg^{2+}, Al^{3+} and Si^{4+} each possess ten electrons, but they are pulled closer to the silicon nucleus (charge 14+) than to the sodium nucleus (11+). Anion radii are larger, owing to the extra electrons they have acquired and to the repulsion between them. O^{2-}, F^-, S^{2-} and Cl^- are larger than all cations except those of the alkali metals and the heavier alkaline earth elements (Box 7.2). Consequently many crystal structures can be visualized as close-packed arrays of large anions, with smaller cations occupying the **interstitial** holes between them.

THE RADIUS RATIO AND ITS APPLICATIONS

When spheres of the same size are packed together, the most compact arrangement consists of a stack of regular planar layers. Every layer has hexagonal symmetry reminiscent of a honeycomb, each sphere being in contact with six others in the same plane. Similar layers can be stacked on top, each sphere nestling in the depression between three spheres in the layer below (Figure 7.2). The spaces between the touching spheres have two kinds of three-dimensional geometry. One type is bounded by the surfaces of four neighbouring spheres. Joining up their centres gennerates a regular **tetrahedron** (Figure 7.2), and in view of their capacity to accommodate small cations such voids are called **tetrahedral sites**.

The second type of hole is bounded by six neighbouring spheres, whose centres lie at the six apexes of a regular **octahedron** (Figure 7.2). Such holes are called **octahedral sites**. In terms of the size of the largest interstitial sphere that each hole can accommodate, octahedral sites are 'bigger' than tetrahedral sites. Both are substantially larger than the cavity between three adjacent spheres in the same layer.

In many ionic crystals the anions are assembled in a more or less close-packed array, the cations occupying some of the tetrahedral and/or octahedral sites between them. The kind of site a given cation occupies is determined by the value of the ratio $r_{cation} : r_{anion}$, known

Box 7.2

Ionic radii of geologically important elements

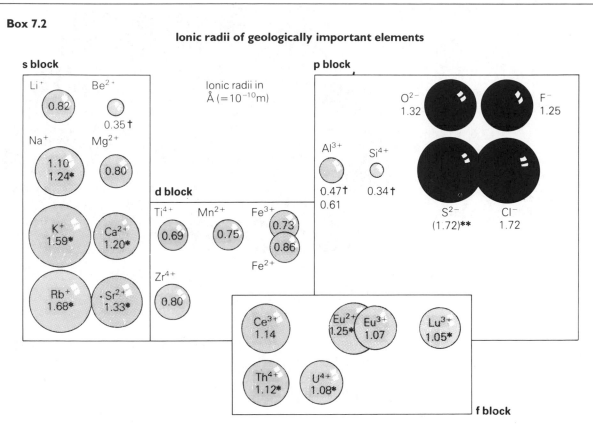

Ionic radii in Å ($=10^{-10}$m)

† Refers to tetrahedral co-ordination
* Refers to 8-fold co-ordination
**The ionic radius of S^{2-} should be used with caution as most metal–sulphide bonds are significantly covalent

Variation with co-ordination number

The equilibrium length of an ionic bond depends on the co-ordination of the ions (due to repulsion between similar ions — Box 7.1). The Al–O distance in analcime $Na(AlSi_2O_6).H_2O$ (in which Al is tetrahedrally coordinated) is 1.8 Å, whereas in jadeite ($NaAlSi_2O_6$, Al octahedrally co-ordinated) it is more than 1.9 Å. It follows that individual ionic radii depend on co-ordination too. This dependence is more significant for cations than anions. The data in the chart above refer to octahedral (6-fold) co-ordination unless otherwise specified. (Data from E. J. W. Whittaker and R. Muntus (1970) *Geochim. Cosmochim. Acta* **34**, 945–56.)

as the **radius ratio**. Using three-dimensional trigonometry, it is not difficult to show that, to fit exactly into an octahedral site between six identical spheres of radius R, a 'cation' sphere must have a radius of $0.414R$. A cation in this position in a crystal is said to have **octahedral co-ordination**. However, the likelihood of a real ion pair having a radius ratio of exactly 0.414 is negligible, so we must consider the effect on the co-ordination number if the radius ratio deviates from this value.

Octahedron with apexes lying at the centres of six surrounding balls, showing the largest sphere that can be accommodated.

Tetrahedron showing the largest sphere that can be accommodated in a tetrahedral site.

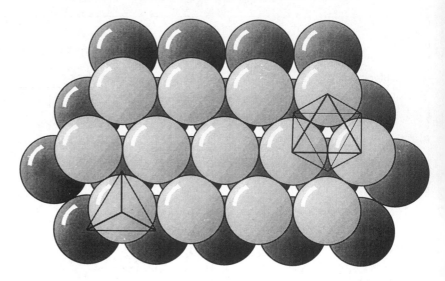

Figure 7.2 Two layers in a close-packed array of spheres. The heavy lines show the co-ordination polyhedra of the tetrahedral and octahedral interstitial voids between the spheres. (After D. McKie and C. McKie (1974) *Crystalline solids*, Nelson, London, with permission.)

A radius ratio of exactly 0.414 allows the 'cation' sphere to touch all of the surrounding spheres at once, maintaining the optimum bond-length (Figure 7.1) with all six anions. This will not be so if the central sphere is smaller than $0.414R$: it will 'rattle' in the hole, and the distance to some of the surrouding ions must exceed the optimum bond-length, violating the energy-minimizing rule above. This unstable situation collapses into a new configuration in which the cation is able to maintain the optimum bond-length with fewer surrounding anions. In practice this means that a cation in this size range will occupy a tetrahedral site in preference to an octahedral one. A radius ratio of less than 0.414 therefore implies **tetrahedral** (four-fold) **co-ordination** of the cation (Table 7.1). For this reason we find silicon (Si^{4+}), with a radius ratio to oxide (O^{2-}) of $0.34/1.32 = 0.25$, in tetrahedral co-ordination in all silicate minerals (which are compounds of various metals with silicon and oxygen).

A radius ratio larger than 0.414 does not prevent the cation maintaining the optimum bond-length with six equidistant anions. The larger cation prevents the anions remaining in contact with each other (they cease to be strictly close-packed), but in view of their mutual repulsion this will not reduce the stability of the structure. Octahedral co-ordination is therefore consistent with radius ratios above 0.414 (Table 7.1). Na^+ is octahedrally co-ordinated in NaCl (radius ratio $1.10/1.72 = 0.64$). Neutrality dictates that Na^+ and Cl^- ions must be equal in number (this is called an 'AB-compound'). It follows that each chloride ion must also be octahedrally co-ordinated by Na^+ ions (Figure 7.3a). This **sodium chloride structure** is shared by the mineral galena (PbS).

Table 7.1 Co-ordination polyhedra.

Range of radius ratio r_{cation}/r_{anion}	Co-ordination number N	Coordination polyhedron
		tetrahedron
0.225 to 0.414 $(1.5^{\frac{1}{2}}-1)$ $(2^{\frac{1}{2}}-1)*$	4	
		octahedron
0.414 to 0.732 $(3^{\frac{1}{2}}-1)$	6	
		cube
0.732 to 1.000 $(4^{\frac{1}{2}}-1)$	8	
>1.0	>12	various

* The upper limit appropriate to co-ordination number N is given by $(N/2)^{\frac{1}{2}}-1$.

In the mineral rutile (one of the polymorphs of TiO_2), the $Ti^{4+}:O^{2-}$ radius ratio is $0.69/1.32 = 0.52$, and accordingly the Ti^{4+} ion is octahedrally co-ordinated by O^{2-} ions. Because there are twice as many O^{2-} ions (the valency difference makes this an 'AB$_2$'-type compound), each oxide ion is found in three-fold co-ordination, occupying the centre of a triangular grouping of Ti^{4+} ions (Figure 7.3c).

The upper limit for octahedral co-ordination is met at a radius ratio of 0.732, at which the cation is sufficiently large to touch eight equidistant neighbours at the same time (Table 7.1). The requirement of maximum co-ordination leads to a new structure in which the anion nuclei lie at the corners of a cube, with the cation enjoying **eight-fold** co-ordination at the centre (**cesium chloride structure**, see Figure 7.3b). The mineral fluorite CaF_2 (Figure. 7.3d) is an example of an AB$_2$

151

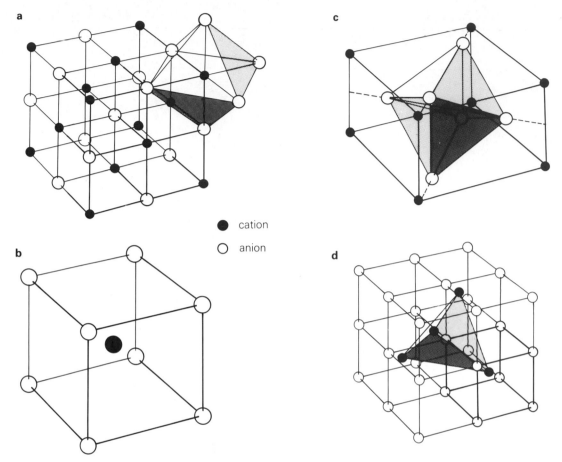

a

c

● cation

○ anion

b

d

Figure 7.3 Structures of simple binary compounds. Ions are depicted at one-tenth of the appropriate size to allow the three-dimensional disposition to be seen.
a. Sodium chloride structure. b. Cesium chloride (CsCl) structure. c. Rutile structure. d. Fluorite structure.

compound with the cation in eight-fold co-ordination. Each fluoride F^- ion lies at the centre of a tetrahedral group of Ca^{2+} ions.

If the radius ratio exceeds 1.0, the co-ordination number rises to 12. Eight-fold and larger sites cannot exist as interstitial sites in a close-packed assemblage of anions, and the presence of large ions like K^+ requires the host crystal to have a more open structure, as for example in feldspars.

Analysing ionic crystal structures as if they were 3D assemblages of hard spheres thus explains why the structure of halite, for example, differs from that of rutile, and also provides a basis for understanding the preferences of major elements for specific sites in silicate minerals (Table 8.2). One must recognize, however, that the ionic model is an idealization, and real crystal structures are complicated by other factors. In many minerals the bonding involves a degree of electron sharing (covalency–see section 7.2) which undermines the assumption that bonding is non-directional, and where transition metals are concerned the presence of d-orbitals introduces further complications. Predictions of crystal structure based on the radius ratio must therefore be seen only as general guidelines.

7.2 The covalent model of bonding

Many substances exhibit chemical bonding between atoms having the same, or very similar, electronegativity. Among them are some of the hardest, most strongly bonded materials known, including diamond, silicon carbide and tungsten carbide. In materials like these, ionic bonding cannot work and a different bonding mechanism must be operating. Although capable of forming extended crystalline structures like diamond, this bonding is also responsible for small discrete molecules like O_2, CH_4, CO_2 and H_2O, the shape of which often indicates a directional type of bond completely foreign to the electrostatically bonded, close-packed materials so far considered. These are the characteristics of **covalent bonding**, which operates through the **sharing** of unpaired electrons between neigbouring atoms.

An **unpaired electron** is one that occupies an orbital on its own. When such singly occupied orbitals in adjacent atoms overlap each other they coalesce to form a **molecular orbital**, allowing the electrons to pass freely between one atom and the other. In this shared state the electrons have a total energy lower than that in either atom individually (Box 7.3), and consequently the atoms have greater stability attached to each other than they had as separate atoms. The greater the degree of overlap, the stronger the attractive force becomes; however, as with ionic bonding (Box 7.1), the tendency for the atoms to continue approaching each other is restricted by a short-range repulsion between the core electrons of the two atoms, and ultimately between their nuclei. The equilibrium covalent bond-length between two atoms can be divided up into the covalent radii of the individual atoms, but the values differ numerically from the corresponding ionic radii.

7.2.1 σ and π bonds

The unpaired electron in a hydrogen atom is found in the 1s orbital, so the coupling in a hydrogen molecule H_2 is due to 1s overlap. The molecular orbital so formed (Box 7.3) is designated 1sσ, and its occupation by two electrons can be symbolized by the electron configuration 1sσ^2. A σ-**bond** is one with cylindrical symmetry about the line joining the two nuclei (Figure 7.4a). σ-bonds can also form when two p-orbitals overlap end-on, or when a p-orbital in one atom overlaps with an s-orbital in another, as in the water molecule (Figure 7.4b). The

Box 7.3

The mechanism of covalent bonding

When two hydrogen atoms come into contact, each electron feels the attraction of a second nucleus. Both electrons modify their standing waves to extend across the two atoms, forming a **molecular orbital** in which the electrons are identified with the H_2 **molecule** instead of with the separate H **atoms**. The wave function of the molecular orbital can be regarded as the sum of the separate atomic wave functions (see diagram). The electron density (ψ^2) in the region between the two nuclei is enhanced (shaded area) at the expense of other parts of the molecule, and **screens** the nuclei from each other. The molecular orbital offers each electron a lower energy than it had in the isolated atom. Any two atoms having valence electrons in such **bonding** orbitals have a lower energy together than separately, in

spite of the inter-nuclear repulsion. For this reason the hydrogen molecule H_2 is more stable than atomic hydrogen.

A bonding orbital accepts two electrons with opposed spins, accommodating the unpaired electron contributed by each of the bonding atoms. The overlap of atomic orbitals that each contain two electrons (as when two helium atoms touch) does not lead to bond formation. The additional two electrons establish a complementary molecular orbital configuration (an **anti-bonding** orbital, with diminished electron density between the nuclei) whose energy is **higher** than the corresponding atomic orbitals. With electrons occupying bonding and anti-bonding orbitals, there is no net energy advantage in forming a molecule. Thus helium, having no unpaired electrons. cannot form a stable He_2 molecule.

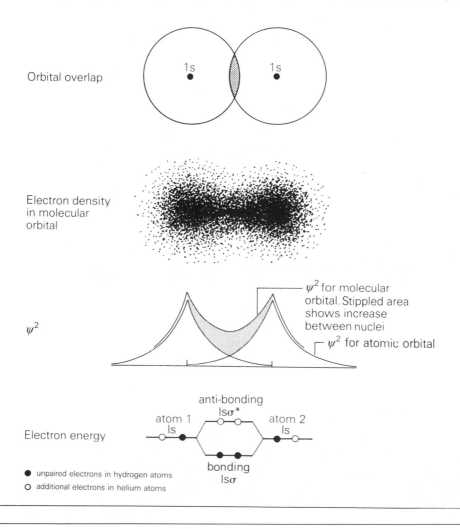

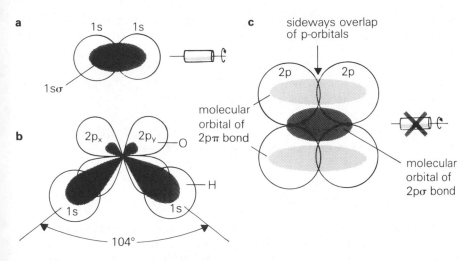

Figure 7.4 Orbital overlap in σ- and π-bonds. The stippled areas show the approximate disposition of the associated bonding molecular orbitals. a. 1sσ bond in the hydrogen molecule H_2. Note the cylindrical symmetry. b. 2p1sσ bonds in the water molecule H_2O. c. 2pσ and 2pπ bonds in a double-bonded molecule.

participation of p-orbitals, with their elongated shape (Figure. 5.5), gives a σ-bond a specific direction, whose orientation in relation to other bonds determines the shape of a multi-atom molecule like H_2O. Note that the two σ-bonds in the water molecule involve separate p-orbitals of the oxygen atom, each of which contributes an unpaired electron; it is not possible for the hydrogen atoms to be attached to opposite lobes of the same p-orbital.

Geometrical constraints prevent the formation of more than one σ-bond between the same pair of atoms. There is however another way in which p-orbitals can link atoms together. Figure 7.4c shows two atoms that are already joined by a 2pσ bond as described above; the σ molecular orbital is shown by the heavy stipple. Suppose each of these atoms has a second unpaired electron in a p-orbital perpendicular to those forming the σ-bond. The two lobes of this orbital lie in the plane of the diagram (large circles), and each can overlap sideways as shown with the corresponding lobe in the other atom, generating two concentrations of electron density (lighter stipple), above and below the σ-bond. Together these constitute the molecular orbital of what is called a **π-bond** between the two atoms. π-bonds form only in conjunction with a σ-bond and are therefore characteristic of molecules containing **double** or **triple bonds**. Such molecules are not cylindrically symmetric. π-bonding occurs in the doubly bonded oxygen molecule (O=O, i.e. O_2), although the configuration is a little more complicated than described above. The triple bond in the nitrogen molecule (N≡N, i.e. N_2) consist of one sigma bond, with two π-bonds in mutually perpendicular planes.

COVALENT CRYSTALS
It is only the lightest elements on the right of the Periodic Table (N, O) that form multiple-bonded, diatomic gas molecules. Heavier elements

in the same groups (V and VI) do not generally form stable multiple bonds; instead they exist as solids consisting of extended, singly bonded molecular structures. The crystal structure of the element sulphur, for example, consists of buckled rings of six or eight sulphur atoms in which each is bonded to two others. These heavier molecules have too much inertia to exist as gases at room temperature.

In diamond (a form of crystalline carbon) and silicon, each atom is bonded to four others in a continuous three-dimensional network (Figure 7.5b). Each perfect diamond crystal can therefore be regarded in a formal sense as a single molecule of carbon.

7.2.2 Molecular shape; hybridization

Methane (CH_4), the chief constituent of natural gas, consists of molecules with a characteristic tetrahedral shape. Four C–H bonds project form the central C atom, as if toward the four 'corners' of a regular tetrahedron, at angles of 109° to one another (Figure 7.5a). It is hard to reconcile this shape, and the existence of four identical bonds, with the electronic configuration of carbon ($1s^2 2s^2 2p^2$) in which the 2s electrons are already paired and therefore not available for bonding. Why is the valency of carbon 4, not 2?

An unpaired valence electron achieves a lower energy in a molecular orbital (if of the 'bonding' type – Box 7.3) than in its own atomic orbital, which is why molecules can be more stable than separate atoms. In forming methane, carbon can **promote** an electron from the 2s orbital into the one 2p orbital that is still vacant: the slight energy disadvantage in doing so (Figure 5.7) is outweighed by having four unpaired electrons available to establish molecular orbitals with hydrogen atoms, rather than just two (Figure 7.5).

The Schrödinger wave analysis of the atom, as well as definig the shape of individuals s- and p-orbitals, offers the possibility of **mixing** their wave funcitons in various proportions to produce **hybrid orbitals** of different geometry. Hydbridization is another consequence of the wave model of the electron. Just as a guitar string can simultaneously vibrate with two or more harmonics of different frequencies (it is the combination of multiple harmonics that gives the instrument its distinctive timbre), so a single atomic electron can adopt more than one waveform. The combined waveform (the 'hybrid' orbital) differs in shape from the individual waveforms from which it is mathematicallly derived. The result of amalgmating the 2s and three 2p orbitals of carbon is called an **sp^3-hybrid**. It consists of four lobes of electron density, each accommodating an unpaired electron, projecting out from the nucleus with tetrahedral symmetry. Each of these lobes forms a separate σ-bond with a hydrogen atom in the methane molecule.

The geometry of the ammonia molecule (NH_3) also results from the sp^3 hybrid. Nitrogen has five electrons in its valence shell (Figure 7.5a). Three of the lobes of an sp^3-hybrid are each occupied by an unpaired

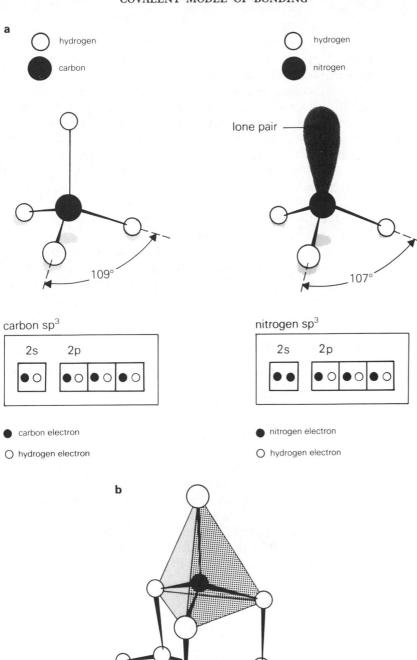

a

hydrogen

carbon

hydrogen

nitrogen

lone pair

109°

107°

carbon sp³

2s	2p		
●○	●○	●○	●○

● carbon electron

○ hydrogen electron

nitrogen sp³

2s	2p		
●●	●○	●○	●○

● nitrogen electron

○ hydrogen electron

b

Figure 7.5 a. The structure of the methane and ammonia molecules. b. The structure of the diamond. The shaded tetrahedron shows the co-ordination of each carbon. In b., atoms are shown at 1/10 true size in relation to inter-nuclear distance. The broad end of each bond projects out of the paper.

electron that forms a σ-bond with an H-atom. The fourth lobe accommodates the remaining two electrons, which being paired are not available for bonding that involves electron sharing (covalent bonding). The electron density of this **lone pair** repels each of the N–H molecular orbitals, leading to a distorted hybrid in which the angle between the bonds is only 107°. In the water molecule, two of the sp³ lobes on the oxygen atom contain lone pairs, and their combined repulsion closes the angle between the O–H bonds still further (to 104°–Figure 7.4b).

The tetrahedral sp³ hybrid can be recognized in the structure of diamond (Figure 7.5b), the unique hardness of which reflects the strong bonds that each carbon atom forms with its four neighbours. The same structure is found in metallic silicon, also a fairly hard material. sp³ geometry is characteristic of single-bonded carbon compounds, like the saturated hydrocarbons (Chapter 9). When carbon forms double bonds, however, it utilizes the alternative **sp²-hybrid**. This has a different geometry, found in the mineral graphite (Box 7.4) and in the benzene molecule (C_6H_6), the basic unit of aromatic carbon compounds (Chapter 9).

7.2.3   The co-ordinate bond

In some circumstances a covalent-like bond can also be established using a lone pair. This arises in co-ordination complexes, which consist of a central metal atom or ion (commonly a transition metal) surrounded by a group of **ligands**, electronegative ions or small molecules (such as NH_3 or HS^-) that possess lone pairs of valence electrons. By overlapping with an empty orbital in the metal atom, each ligand lone pair forms a bonding molecular orbital which can accommodate the two electrons at a lower energy. This **co-ordinate bond** is a variety of covalent bond in which both electrons are supplied by one of the participating atoms (the ligand) instead of one electron being supplied by each of them. Such complexes are stable because each electron enjoys the attraction of two nuclei (cf. Box 7.3). Current chemical thinking views the formation of a co-ordinate bond as a particular type of acid–base reaction (Box 7.5).

Co-ordination complexes markedly increase the solubility of many ore metals in saline aqueous solutions such as hydrothermal fluids (Chapter 4).

7.2.4   Metals and semiconductors

In a crystal of pure copper, no electronegativity difference exists between neighbouring atoms, and we expect the Cu–Cu bonding to be covalent. Yet the characteristic properties of metals – lustre, opaqueness, electrical and thermal conductivity, and ductility – make them quite different from the covalent compounds considered so far.

Box 7.4

Case study: bonding in graphite

Graphite is a soft, greasy, black, lustrous mineral, widely used as a lubricant and electrical conductor. It is impossible to imagine a greater contrast between these properties and those of the other carbon polymorph, diamond. Why are they so different?

The diagram shows that graphite, unlike diamond, is made up of sheets. Each sheet is a continuous network of interconnected hexagonal rings, in which every carbon atom is bonded to three equidistant neighbours, the bonds sticking out symmetrically at 120° to each other. This 'trigonal planar' bonding geometry is the signature of the sp^2-hybrid, formed by combining the 2s orbital of carbon with *two* of the three 2p orbitals. The three co-planar sp^2 lobes establish σ-bonds with neighbouring atoms.

The C–C bond length in the graphite sheets is 1.42 Å, a little shorter than the C–C distance in diamond (1.54 Å). As stronger bonds pull atoms closer, this suggests that the intra-sheet bonding in graphite is stronger than the bonding in diamond. The explanation lies with the fourth unpaired electron of carbon, excluded from the sp^2-hybrid. This occupies a p-orbital projecting perpendicular to the sheet, and it allows π-bonds to be formed between neighbouring atoms. reinforcing the intra-sheet σ-bond network.

In classical bonding theory, this π-bond would be added to one of the three σ-bonds, transforming it into a double bond with the other two remaining single. The diagram shows there are three altermative ways of distributing these double bonds throughout the sheet so that every atom is involved in four bonds. In the wave-mechanical interpretation, however, the real situation is a combination of all three configurations: the p-electron in each atom can be involved in partial π-bonds with all three neighbours at once. Consequently the π-bond 'sausages' of electron density above and below the plane (Figure 7.4c) are not localized between two specific atoms but are spread thinly all round the hexagon and indeed throughout the sheet. The π-electrons are **delocalized** in interconnecting molecular orbitals resembling sheets of chain-link fencing, in which the energy structure resembles that of a metal (Box 7.6) and allows them to migrate across the entire sheet if an electric field is applied. Thus graphite is a good electrical conductor in directions parallel to the sheets. This conducting behaviour is also responsible for the opaque, metallic appearance of graphite.

Intra-layer bonding in graphite ties up all four valence units of carbon. The adhesion of one sheet to another (inter-layer bonding) is achieved through a much weaker attraction called the **van der Waals interaction**, described later in the chapter, that fails to prevent sheets slipping easily over each other. The extreme softness of graphite and the large inter-layer spacing (3.35 Å) indicate how feeble the inter-layer force is. As there is no inter-communication of electron orbitals between sheets, graphite acts as an insulator in directions perpendicular to the sheets.

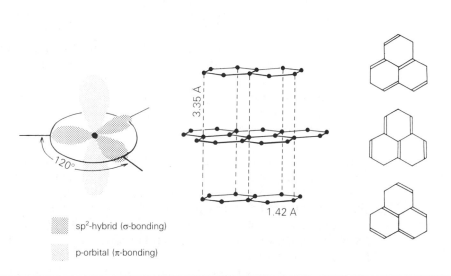

3.35 Å

1.42 A

120°

sp²-hybrid (σ-bonding)

p-orbital (π-bonding)

Box 7.5

Lewis acids and bases

In 1923 G. N. Lewis widened the concept of acids and bases to encompass systems in which H^+ ions (upon which the traditional notion of acids and bases is founded – Appendix C) are not available, such as silicate melts. Lewis' ideas also have particular relevance to co-ordination complexes, in which bonding involves the sharing of a 'lone pair' donated by one of the participating atoms. Lewis defined a base as 'an atom or molecule capable of donating an electron pair to a bond' whereas a 'Lewis acid' is an atom or molecule that can accept a lone pair. The Lewis definition of 'acid' embraces the traditional 'H^+ donor' viewpoint (since H^+ can readily attach to a lone pair, forming a covalent bond) but has much wider application.

How can we apply the Lewis concept to co-ordination complexes? As we saw on p. 100, the co-ordination complex $Cu(HS)_3^{2-}$ is believed to play an important part in the low-temperature hydrothermal transport of Cu. The electron configuration of the Cu^+ ion is $[Ar]\ 4s^0 3d^{10} 4p^0$. The complex forms because a lone pair of electrons on each HS^- ion overlaps with a vacant Cu orbital (4s or 4p), forming a molecular orbital that allows the electron pair to be associated with two atoms (S and Cu). Here each HS^- ion is acting as a *Lewis base* (electron-pair donor) and the Cu^+ ion is the *Lewis acid*.

The Lewis approach provides valuable insights into the chemistry of silicate melts, and helps us to understand why, for example, some metals prefer to be associated with sulphide minerals whereas others have more affinity with silicates (Box 9.7).

The explanation of these differences lies in the molecular orbitals by which a metallic crystal is held together.

When a hydrogen atom forms a covalent bond, the valence shell becomes in effect fully occupied. The electron configuration in the hydrogen molecule ($1s\sigma^2$), for example, is equivalent from the atom's point of view to the configuration of helium. A sulphur atom participating in two bonds similarly attains a noble-gas configuration (Ar). The picture in a metal differs from these elements in two ways:

(a) Metals have lower electronegativities and ionization energies than hydrogen and sulphur. Electrons in their valence shells are more loosely held.

(b) The formation of covalent bonds does not lead to a noble-gas configuration in a metal atom. Vacant energy levels remain available in the metal's valence shell.

In any covalent crystal, the overlap of valence orbitals between neighbouring atoms leads to a system of molecular orbitals extending throughout the crystal. The electrons occupying them are nominally shared by all of the atoms present. The energy levels available to these electrons are grouped in bands (Box 7.6), and the physical properties of a crystal depend on how these bands are arranged in energy space and how they are populated.

Figure 7.6 shows three alternative situations. In diamond – and in

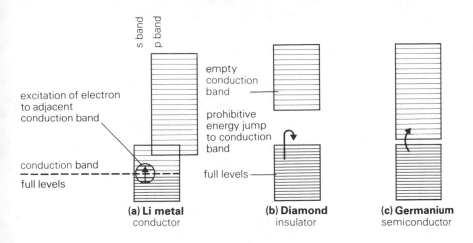

Figure 7.6 Energy bands in (a) a conductor, (b) an insulator and (c) a semiconductor.

sulphur, not shown – the valence electrons completely fill the lower band (equivalent to filled valence shells in all atoms). Each electron has a molecular wave function that confines it to a particular location in the crystal. In order to migrate it must find a vacancy in another molecular orbital, but no vacancies exist except in the upper band at unattainable energy levels. The essential characteristic of an **insulator** is therefore a filled band, from which vacant higher levels are separated by a large gap.

In a metal, valence electrons can fill only the lower part of the band. Upper levels are vacant and, together with other overlapping bands which are readily accessible (Figure 7.6), serve as a **conduction band** that enables electrons to migrate to new positions in the crystal. Electrons are not stable in these higher levels, but thermal excitation from the filled levels below is sufficient to ensure that a proportion of such levels is always occupied. A voltage applied across the crystal will produce the net flow of electrons that constitute an electric current. Similar energy bands in the π-bonding network lie behind the electrical conductivity of graphite (Box 7.4).

One can visualize a metal as a regular array of 'cations' (though in this context ionic radii are not applicable) immersed in a fluid of mobile, **delocalized** valence electrons. Owing to the mobility of electrons, the bonding in metals is essentially non-directional. Most metals have close-packed atomic structures determined by the stacking rules for spheres of equal size. Twelve-fold co-ordination is therefore the commonest configuration, although a number of metals show eight-fold co-ordination instead.

Hydrogen was cited above as an example of a non-metallic element, as it occurs in the elemental state on Earth solely in the diatomic form H_2. Experiments at extremely high pressure have shown, however, that highly compressed hydrogen can develop the electronic band structure characteristic of a metal (Box 7.6), and it is believed to exist in this

161

Box 7.6

The energy-level structure of lithium metal

Box 7.3 showed that 1s orbital overlap between hydrogen atoms generates two molecular orbitals: a bonding orbital (1sσ), having a lower energy than the atomic orbital; and an anti-bonding orbital (denoted 1sσ*), with a higher energy. The same happens when two lithium atoms form a bond.

Because these 2sσ molecular orbitals extend across the two atoms, the electrons in them are common to both. When three Li atoms are close enough for overlap, to occur, they share three electrons and form three molecuar orbitals with three distinct energy levels. Four atoms form four orbitals; five atoms form five; and n atoms n orbitals. In a crystal of appreciable size, the value of n is effectively infinite, resulting in a virtually continuous energy **band**. As it originates from interaction between s-orbitals, it is called the

s-band. The n valence electrons occupy the lower (bonding) half. These lower levels represent fixed electron states, each associated with a particular location in the crystal.

An electron migrates by transferring to a new molecular orbital. The only vacant ones are those with energies in the upper (anti-bonding) part of the band, and promotion of electrons to these accessible upper levels explains the electrical conductivity of metals (and the optical properties derived from it – Chapter 8). Electrons may also have access to higher bands; in Li the hypothetical overlap of empty p-orbitals makes available a 'p-band', partly interleaved with the s-band (Figure 7.6). These vancant levels constitute the **conduction band**.

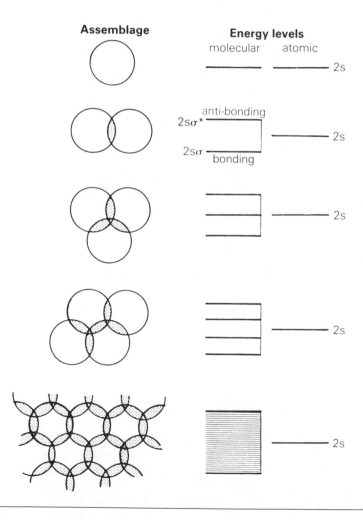

metallic form in the deep interiors of Jupiter and Saturn (Open University, 1994, cited in Chapter 10). The strong magnetic fields of Jupiter and Saturn are attributable to electric currents flowing in this metallic hydrogen fluid.

SEMICONDUCTORS

Germanium (Figure 7.6c) illustrates a technologically important intermediate case, the semiconductor. If a small gap exists between a filled band and an empty conduction band, the material will behave as an insulator except when activated by an energy impulse from outside. One of the uses of germanium, for example, is as a detector of γ-rays (Box 6.3), such as those emitted by certain trace elements in geological materials when irradiated by neutrons in a nuclear reactor (an analytical technique called 'neutron activation analysis'). Silicon is used in a similar way as an X-ray detector in the electron microprobe (Box 6.4). A Ge (or Si) crystal, even with several hundred volts applied across it, will conduct no appreciable current except when it absorbs a γ-ray (or X-ray) photon. The photon energy is sufficient to promote a number of electrons temporarily into the conduction band, leaving the equivalent number of 'holes' in the valence band. This event produces a short current pulse before the crystal relaxes back to its insulating condition. As the number of electrons promoted is proportional to the photon energy, the amplitude of output pulse will vary in proportion to the quantum energy of the photon detected. Electronically sorting the pulses according to pulse height therefore provides a simple and economical means of isolating the various components of a complex γ-ray or X-ray spectrum; this is the basis of energy-dispersive ('ED') spectrometry (Box 6.4). The intensity of each spectral 'line' is obtained by counting the number of pulses that are recorded within the appropriate pulse-height interval.

The energy gap between filled and conduction bands in a semiconductor can be deliberately engineered for specific applications by 'doping' the crystal surface with other elements, notably those of Groups IIIb and Vb; the doping elements introduce additional energy levels into the gap.

7.3   Bonding in minerals

We have seen that ionic bonding develops between elements of very different electronegativity, whereas covalent bonds are characteristic of materials, including pure elements, in which the electronegativity contrast is slight. Most of the compounds we meet as minerals fall between these two extremes, so we have to consider what sort of chemical bonding operates between elements that show moderate differences in electronegativity. We shall investigate this intermediate domain by considering how real bonds deviate from the idealized ionic and covalent models.

7.3.1 Anion polarizability; non-ideal ionic bonds

We have postulated an ideal ionic bond in which an electron is completely transferred from a donor atom (which becomes a cation) to a host atom (which becomes an anion). Ideally the electron should be associated solely with the anion nucleus, and its spatial distribution should be symmetrical about it. In reality, the net positive charge on each neighbouring cation acts as a competing focus of attraction and to a small extent the electron density of the anion is concentrated in the region between the nuclei. In other words, a degree of electron sharing (partial covalency) occurs in any real ionic bond. The ideal ionic bond does not exist in nature.

The degree of polarization of the anion is obviously greater if the cation is highly charged (Mg^{2+} as compared to Na^+, for example). The polarizing field of the cation is also more intense if the cation is small, because its charge can then be brought closer to the anion. The ratio of the cation's charge to its radius, which is called the **ionic potential** of the cation, is therefore a measure of its power to polarize an anion and recover a fraction of the excess electron density. Large singly charged cations like K^+ possess little polarizing power and form the 'purest' ionic bonds. Small multiply charged cations like Al^{3+} and Si^{4+} are highly polarizing, and their compounds can be considered as partially covalent (Figure. 7.7).

A cation's ionic potential – a property related to the element's electronegativity – is a useful guide to its behaviour in molten and crystalline silicates (Chapter 8) and in aqueous solution.

7.3.2 Polarization of a covalent bond; ionicity

In an ideal covalent bond, two valence electrons are shared equally between two atoms. The two atoms must have equal power to attract electron density to keep it positioned symmetrically between them. If they differ even slightly in electronegativity, electron density will be gathered disproportionately around the more electronegative atom, giving it a slight negative net charge (and leaving a complementary positive charge on the other atom). Such 'polarization' of the covalent bond is equivalent to transferring a fraction of an electron from one atom to the other, introducing a degree of ionic character or 'ionicity' into the bond.

Pauling pointed out that a continuous progression in bond type must exist between purely covalent and predominantly ionic bonds. The character of an intermediate type of bond can be quantified in terms of the **proportion of ionic character** present, expressed as a percentage, which Pauling correlated with the difference in electrone-

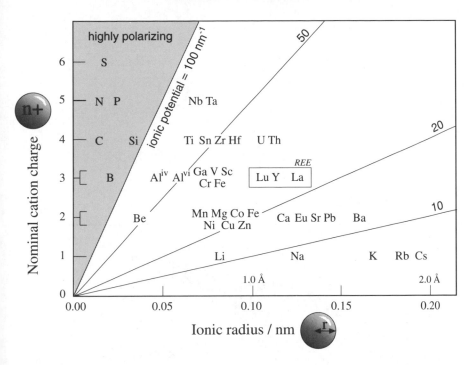

Figure 7.7 Ionic potentials of common 'cations'. 'REE' refers to the rare earth elements La (Z = 57) to Lu (Z = 71) – see Chapter 6. Yttrium (Y) has similar properties.

gativity between the participating atoms, as shown in Figure 7.8a. Though not an exact relationship (and therefore shown as a band in the figure), the correlation offers a valuable insight into the properties of minerals.

7.3.3 Bonding in silicates

One particularly important bond in view of its rôle in silicate structure (Chapter 8) is the Si–O bond, which appears from Figure 7.8a to combine ionic and covalent bonding in more or less equal amounts. The small ionic radius of the Si^{4+} 'cation' (0.34 Å, Box 7.2) is consistent with its occupation of tetrahedral sites in all silicate minerals. In view of its high charge and small radius, however, the Si^{4+} cation must be highly polarizing and its distortion of the oxygen ion introduces an appreciable degree of covalency into the Si–O bond. The concept of an Si^{4+} cation is therefore an approximation to be used with caution: the charge residing on a silicon atom in a silicate structure actually approximates to 2+ rather than 4+. Silicon, like carbon, uses an sp^3-hybrid in forming covalent bonds, and tetrahedral co-ordination is therefore as much a reflection of covalent bonding between Si and O as of ionic bonding. The relative covalency of the Si–O bond accounts for the structural coherency of the chain, sheet and framework skeletons of many silicate minerals (Chapter 8). Like phosphorus and sulphur, Si shows little tendency to form double or π-bonds. Unlike C, it never forms an sp^2-hybrid.

165

Figure 7.8 a. The correlation of percentage ionic character in a bond with the difference in electronegativity, after Pauling. The approximate status of geologically relevant bonds is shown. b. Oxide, sulphide and elemental bond types shown on a solid figure whose axes are ionicity, mean electronegativity and electronegativity difference. The dashed line links the bonding of iron in oxide (silicate), sulphide and metal states.

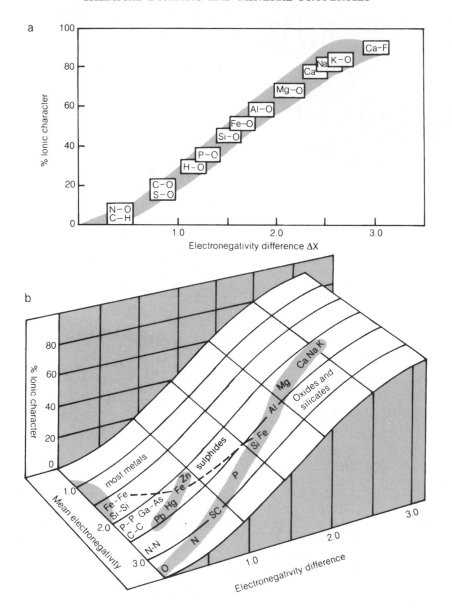

The other chemical bonds operating in silicates are all more ionic in character than the Si–O bond. Al^{3+} is a small ion with a relatively high charge, but its ionic potential is barely more than half that of Si^{4+} (Figure 7.7) and the Al–O bond is regarded as being nearly 60% ionic (Figure 7.8a). Mg–O is about 65% ionic and Ca–O, Na–O and K–O are all more than 75% ionic. For these elements the ionic model provides the most appropriate interpretation of co-ordination and crystal structure. The same of course applies to the halides of these elements, such as the minerals halite (NaCl) and fluorite (CaF_2).

7.3.4 Oxy anions

Oxygen forms bonds with phosphorus, carbon, sulphur and nitrogen which are distinctly more covalent than Si–O (Figure 7.8a). Thus in a mineral like calcite we see the different behaviour of two types of chemical bond:

(a) Ca^{2+} and CO_3^{2-} are attracted to each other by ionic bonds which, being electrostatic, break down when calcite is dissolved in a polar solvent, releasing separate calcium and carbonate ions stabilized by electrostatic association with surrounding water molecules (hydration – Box 4.1).

(b) The bonds between C and the three O atoms in the carbonate ion are largely covalent, and the anion retains its identity and structure whether in solution or in crystalline form. When calcite is heated to high temperatures (about 900 °C), the carbonate ion breaks down, yielding carbon dioxide gas.

Similar behaviour is seen with other 'oxy anion' compounds such as phosphate, nitrate and sulphate. They are related in a formal sense to corresponding oxy acids (carbonic, phosphoric, nitric and sulphuric acids). Note that OH^- is another oxy anion, important in minerals like brucite $(Mg(OH)_2)$ and mica.

7.3.5 Pure elements, alloys and sulphides

Crystals of pure elements, whose atoms have uniform electronegativity, will plot at the origin in Figure 7.8a. This group includes, as we have seen, a wide range of behaviour, from insulators like diamond and crystalline sulphur, through semiconductors, to fully conducting metals. These differences can be appreciated by adding the mean electronegativity as a third dimension to the diagram shown in Figure 7.8a, the curving line of which is transformed into the curved surface shown in Figure 7.8b. The oxide bonds form a diagonal trend on this surface. At the foot of this trend lies the O–O bond, one of a group of non-metal bonds extending along the edge of the surface to phosphorus (P–P). The next element, silicon, is a semiconductor. Beyond Si lies a field in which all of the common metals plot. It extends some distance to the right to include alloys like brass (Cu–Zn).

The metallic appearance of many sulphides, illustrated by the popular reference to pyrite (FeS_2) as 'fools' gold', reflects their intermediate position between the oxide and metallic fields in Figure 7.8b. The dashed line shows that the Fe–S bond is intermediate in character between the Fe–O and Fe–Fe bonds. The structural reasons for sub-metallic behaviour in sulphides are considered in Box 9.8.

7.4 Other types of atomic and molecular interaction

We have seen that ionic, covalent and metallic bonding form a unified spectrum of chemical interaction between atoms that possess incomplete valence shells. These bonding mechanisms, considered separately or in combination, account for most of the diversity of appearance, structure and behaviour that we see in minerals. There are nevertheless some associations between atoms and molecules that cannot be explained in these terms.

Various types of weak electrostatic attraction operate between all molecules, whether or not they possess electric charge. Many molecules, owing to internal electronegativity differences, are slightly polarized; and the electrical field associated with such **dipoles** (Figure 7.9a) makes them exert, and be susceptible to, electrostatic forces. A dipole can be attracted to an ion, or to another dipole. It may, by means of its electric field, even **induce** an unpolarized molecule to become a dipole, and thereby attract it. Such dipole interactions, though much weaker than ionic bonding (an 'ion–ion interaction'), have great mineralogical significance.

7.4.1 Ion–dipole interactions; hydration

The shared electron density in an O–H bond is partially concentrated at the oxygen end, owing to the higher electronegativity of oxygen. This polarization, symbolized by a partial positive charge $\frac{1}{2}\delta+$ on each hydrogen atom and a partial negative charge $\delta-$ on the oxygen (Figure 7.9b), results in the water molecule as a whole acting as a dipole.

In aqueous NaCl solution, each Na^+ ion is surrounded by a diffuse

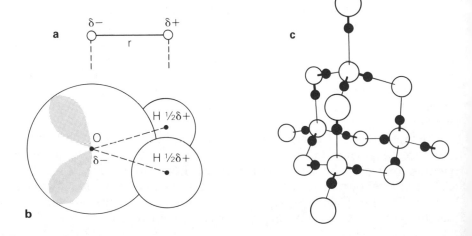

Figure 7.9 Dipole interactions. a. A dipole consists of equal and opposite charges, a fixed distance apart. b. The water molecule as a dipole. c. The structure of ice. Atoms are shown at (1/10) of the appropriate size. Heavy lines are covalent bonds, thin lines are hydrogen bonds. Note the difference in O=H and O–H bond lengths.

blanket of water molecules whose negative 'poles' are attracted toward the Na^+ cation, and around each Cl^- ion is a similar layer with postive poles aligned towards the negative ion. This phenomenon of **hydration** (Box 4.1) is an **ion–dipole interaction**. The water molecule possesses no net charge, but is attracted towards an ion because its oppositely charged (attracted) pole gets closer to the ion than the similarly charged (repelled) pole. The attractive force on the molecule is therefore stronger than the repulsive force.

7.4.2 Dipole–dipole interactions; hydrogen bonding

The unique physical and chemical properties of water (Box 4.1) suggest that some kind of attraction exists between the molecules in water and ice. This attraction, known as the hydrogen bond, arises from the electrostatic interaction between one water molecule and its neighbours, a **dipole–dipole interaction**. This is seen most clearly in the regular structure of ice (Figure 7.9c), in which each hydrogen atom in the molecule is attracted electrostatically toward the oxygen atom in a neighbouring molecule. The attraction is maximized if the geometry allows the hydrogen to associate with one of the lone pairs on the oxygen atom. Owing to the nearly tetrahedral disposition of bonding orbitals and lone pairs in the oxygen atom, this leads not to a close-packed structure but to a three-dimensional network of molecules analogous to the structure of diamond (Figure 7.9c).

This ordered structure breaks down during melting, but about half of the hydrogen bonding is preserved in liquid water, giving rise to its high viscosity, surface tension and boiling point (Box 7.7). Because the molecules are able to pack closer together in the liquid state than in the ordered crystalline framework of ice, water exhibits a slight increase in density on melting, and a corresponding increase in volume on freezing, another unique property which we take for granted (Box 2.2).

Hydrogen is the only element capable of 'bridging' between molecules in this way, a capacity it owes to the fact that its nucleus is not shielded by any inner electron shells: it is exposed to the full attraction of an approaching oxygen lone pair. Hydrogen bonding has considerable structural significance not only in water and ice but in all sorts of compounds in which hydrogen is associated with a strongly electronegative element like oxygen and nitrogen. Hydrogen bonding is largely responsible for the cohesion between layers in minerals like kaolinite, for example. It plays a dominant rôle in most biological processes (Gribbin, 1985), determining for instance the regular arrangement of polypeptide chains in protein molecules (Chapter 9), and cross-linking the double helix in DNA.

7.4.3 Induced-dipole (van der Waals) interactions

The electrostatic field of a dipole can induce a polarization in an atom that is not itself intrinsically polarized. The induced dipole is aligned in

Box 7.7

Estimating the strength of chemical bonds

The strength of a chemical bond is expressed in terms of the energy required to break it. This is usually quoted in the form of the molar enthalpy change for a specific hypothetical dissociation reaction such as:

$$H_2 \rightarrow H + H$$

This enthalpy change, known as the **dissociation energy** of hydrogen, indicates how much energy (in $kJ\,mol^{-1}$) is required to split every molecule in one mole of hydrogen gas into two separate atoms. Some relevant dissociation energies are given in the table.

Type of bonding	Reaction			Value $kJ\,mol^{-1}$	Type of bonding	Reaction			Value $kJ\,mol^{-1}$
ionic	NaCl *crystal*	$\rightarrow$	Na$^+$ *gas* + Cl$^-$ *gas*	767		Pb	$\rightarrow$	Pb	194
						Ni	$\rightarrow$	Ni	423
	NaCl *crystal*	$\rightarrow$	Na *gas* + Cl *gas*	641	hydration	Na$^+$ *gas*	$\rightarrow$	Na$^+$ *aq*	405
convalent	C *diamond*	$\rightarrow$	C *gas*	723	hydrogen bonding	H$_2$O *ice*	$\rightarrow$	H$_2$O *gas*	47
	H$_2$O *gas*	$\rightarrow$	2H + O *gas*	2 × 466	van der Waals	Ar *solid*	$\rightarrow$	Ar *gas*	6.3
metallic	Na *crystal*	$\rightarrow$	Na *gas*	109					

Detailed interpretation of these numbers (for example the estimation of the energy of an individual C–C bond) is fraught with pitfalls. We shall simply note the following generalizations:

(a) Covalent and ionic bond energies fall within a similar range. Energies of single covalent bonds are generally betwen 200 and $500\,kJ\,mol^{-1}$. Note the similarity to the activation energies noted for silicate reactions in Chapter 3 (Figures 3.6 and 3.8).

(b) Metals have similar values, those for soft metals being lower than those for hard metals (Na < Pb < Ni).

(c) Hydrogen-bond energy is about a factor of ten less than for covalent or ionic bonds. For an individual hydrogen bond the energy is around $25\,kJ\,mol^{-1}$.

(d) Van der Waals energies are somewhat lower again than hydrogen bonds.

In the absence of quantitative data, one can make a crude estimate of the energy of interaction between molecules in a covalent substance from its boiling point. This is the temperature at which the energy of the molecules' thermal vibration (which is roughly $\frac{3}{2}RT$ per mole) exceeds the energy of cohesion holding the molecules together.

Elements and compounds of low molecular weight that are gaseous at room temperature (He, Ar, CO_2, N_2) have weak intermolecular forces, but in a solid or liquid material such as water the forces of cohesion are quite strong.

such a way that it is attracted to the original dipole. Curiously this attraction can operate even between two atoms that have no permanent dipoles. To see how, we have to look a little closer at the oscillation of the electron standing wave in the atom (Chapter 5).

We have gained some understanding of the nature of electron orbitals in atoms and molecules by comparing them with standing waves on a vibrating string. Denied by the Uncertainty Principle the opportunity to follow the electron along a precise course in time, we have considered the orbital as a time-independent envelope delineating the electron's spatial domain, analogous to the fixed envelope just visible in a vigorously plucked guitar string. Just as the string oscillates back and forth within its envelope, too fast for us to see, so the electron must also be vibrating within its own orbital. In support of this view one may note that wave mechanics associates an angular momentum with the trapped electron (in orbitals other than s-orbitals), suggesting that the electron, in its own obscure way, does really 'travel round the nucleus'.

The consequence of this oscillation is that if we were able to freeze an atom or molecule at some instant, we would find that electron density in each orbital was not uniformly and symmetrically distributed, but momentarily concentrated in one part or another, making the atom or molecule for that instant a dipole. At each instant the dipole generates an electric field that can polarize neighbouring atoms or molecules in concert with its own oscillation, and attract them closer. Although the electric fields of these synchronous dipoles average out to zero over a period of time, the inter-molecular attraction they cause does not.

The induced-dipole or **van der Waals interaction** is responsible for a weak attraction operating between any pair of atoms or molecules that are sufficiently close (a few Å), though it can be detected only when other inter-atomic forces are absent. Van der Waals forces are responsible for holding the sheets together in crystals of graphite, and no doubt contribute to the cohesion of soft sheet silicates like pyrophyllite and talc. The low hardness of graphite (1–2) and the large inter-layer distance (Box 7.4) show how feeble the attraction is. The 'bond energy' for the van der Waals interaction is typically two orders of magnitude less than for ionic and covalent bonding (Box 7.6). The interaction explains why all substances, even noble gases, form crystals at sufficiently low temperatures (neon for instance melts at 24.6 K and argon at 84 K). These low melting temperatures signify a force so weak that even very slight thermal vibration is sufficient to overcome it (Box 7.7).

7.5 Bibliography

Bloss, F. D. (1971) *Crystallography and crystal chemistry*, Holt, Rinehart and Winston, New York.

Fyfe, W. S. (1964) *Geochemistry of solids*, McGraw-Hill, New York.

Gribbin, J. (1985) The bond of life. *New Scientist*, 19/26 December 1985.

Henderson, P. (1982) *Inorganic geochemistry*, Pergamon, Oxford.
Shriver, D. F., Atkins, P. W. and Langford C. H. (1994) *Inorganic chemistry*, 2nd edn, Oxford University Press, Oxford.

7.6 Exercises

1 Predict the co-ordination numbers of the following ions in a silicate crystal (Box 7.2):

Si^{4+}, Al^{3+}, Ti^{4+}, Fe^{3+}, Fe^{2+}, Mg^{2+}, Ca^{2+}, Na^+, K^+

2 Explain the difference in ionic radius (Box 7.2) between Fe^{2+} and Fe^{3+} and between Eu^{2+}, and Eu^{3+}. Calculate the ionic potentials of the ions. What effect does the oxidation state have on the covalency of bonding of these elements?

3 Identify the type of bonding between the following pairs of atoms in the solid state. How does it influence the properties of the elements at room temperature? (Use Figure 6.3. Holmium (Ho) is a lanthanide metal.)

$$He-He \quad Ho-Ho \quad Ge-Ge$$

4 Discuss the bonding in the following minerals:

KCl	(sylvite)
TiO_2	(rutile)
MoS_2	(molybdenite)
NiAs	(niccolite)
$CaSO_4$	(anhydrite)
$CaSO_4.2H_2O$	(gypsum)

8

Silicate crystals and melts

The majority of rock-forming minerals are silicates, compounds in which metals are combined with silicon and oxygen. In this chapter we consider how the chemical structure of these compounds, particularly the nature of the bonding, determines the familiar morphological and physical properties of the silicate minerals. A minimal knowledge of crystallography will be needed.

The relative convalency of the Si–O bond gives silicon a fundamental structural rôle as the principal **network-forming** element in silicate crystals and melts, establishing the structural skeleton upon which their properties depend. In this respect silicon (with phosphorus and to some extent aluminium) contrasts with the more ionic constituents of silicates like Mg^{2+} and K^+, which are **network-modifying** elements that influence structure only because they restrict they way in which Si–O networks are stacked together.

8.1   Silicate polymers

Whether silicates are examined from the ionic or covalent viewpoint, the behaviour of silicon is the same: it lies at the centre of a tetrahedral group of four oxygen atoms or ions. In structural terms, the **SiO_4 tetrahedron** (Figure 8.1) is the basic building brick, from which all silicate crystals and melts are constructed. But the silicon atom itself can satisfy only half of the bonding capacity of its four oxygen neighbours (four bonds out of a total of eight). How are the remaining oxygen bonds used?

If, when the silicate crystallizes, there is a high concentration of an electropositive element like magnesium (which forms a **basic oxide** – Box 8.1), relatively ionic bonds are likely to be established between each 'tetrahedral' oxygen and a nearby Mg^{2+} ion, the SiO_4 group acquiring in the process an overall negative charge (nominally 4−).

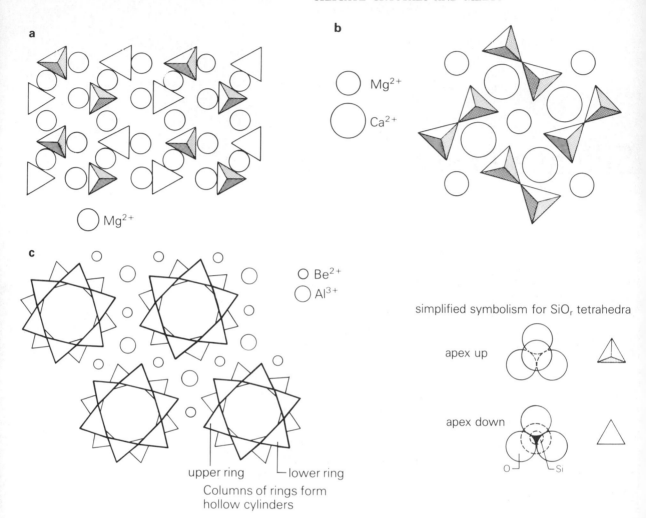

Figure 8.1 Simplified structures of silicate minerals, with cations drawn approximately to scale. SiO₄ groups are shown as bare coordination tetrahedra to clarify the structure. a. Olivine, viewed down the *a* crystallographic axis; b. melilite; and c. beryl.

This leads to the chemically simple crystal structure of **olivine** (in particular the end-member forsterite, Mg_2SiO_4) in which Mg^{2+} ions alternate with SiO_4^{4-} tetrahedra. There is no direct Si–O bonding between one tetrahedron and its neighbour, and the cohesion of the crystal as a whole arises from ionic bonding between Mg^{2+} cations and SiO_4^{4-} anions: covalently bonded SiO_4 'bricks' are cemented together by an ionic 'mortar' of Mg^{2+} ions.

If cations Mg^{2+} are scarce, on the other hand, oxygen atoms are more likely to bond directly to two silicon atoms, forming a relatively

Box 8.1

Acidic and basic oxides

The composition of any silicate material can be represented as a combination of various metallic oxides such as MgO and K_2O combined with silicon dioxide (silica). Forsterite, for example, can be regarded as the result of reacting two molecules of MgO with one of SiO_2:

$$2MgO + SiO_2 \rightarrow Mg_2SiO_4$$

The character of a silicate depends on the proportions in which different oxides combine. They can be subdivided into three types:

Basic oxides
When an **ionic oxide** dissolves in water, the O^{2-} ion released reacts with a water molecule to form two hydroxyl ions:

$$Na_2O + H_2O \rightarrow 2Na^+ + (O^{2-} + H_2O) \rightarrow 2Na^+ + 2OH^-$$

The production of OH^- ions removes free H^+ ions from the solution, increasing its pH and making it **basic** (Appendix C). For this reason the oxides of the electropositive metals are described as **basic oxides** (Figure 9.4).

Acidic oxides
A covalent oxide dissolved in water, instead of producing OH^- ions, will remove them from solution:

$$CO_2 + (OH^- + H^+) \rightarrow HCO_3^- + H^+$$

(cf. Equations 4.14 and 4.16) resulting in an acidic solution. Similar reactions involving the oxides of sulphur produced by coal-burning power stations are environmentally important in leading to **acid rain** (see O'Neill, 1992, referenced in Chapter 9). In the context of silicates, this category of **acidic oxides** includes the oxides of silicon and phosphorus (Figure 9.4).

Amphoteric oxides
Amphoteric oxides and hydroxides share the property of being able to behave as acids (if reacting with a strong base) and bases (with a strong acid):

$$\begin{array}{cccccc}
Al(OH)_3 & 3HCl & & AlCl_3 & & 3H_2O \\
base & + \quad acid & \rightarrow & salt & + & water
\end{array}$$

$$\begin{array}{cccccc}
Al(OH)_3 & + \quad NaOH & \rightarrow & NaAlO_2 & + & 2H_2O \\
acid & base & & salt & & water \\
& & & (sodium & & \\
& & & aluminate) & &
\end{array}$$

Other elements forming amphoteric oxides are shown in Figure 9.4.

'covalent' bridge between them. Taken to the limit, this can lead to every oxygen atom being shared between neighbouring tetrahedra, extending the Si–O bonding to form a three-dimensional network of connected tetrahedra throughout the crystal. This is the situation we find in the mineral quartz. Because each oxygen is structurally part of two SiO_4 tetrahedra, half as many oxygen atoms are needed to fulfil the co-ordination requirements of silicon. Thus quartz, although in a formal sense built of SiO_4 structural units, has the formula SiO_2.

The formation of extended Si–O networks in silicates is called **polymerization**. Whereas organic **polymers** consist of chains and rings of carbon atoms linked directly to each other (–C–C–C–: Chapter 9), the linkage in silicates is always through oxygen atoms (–Si–O–Si–O–Si–). The degree of Si–O polymerization in a silicate structure is conveniently enumerated by the **number of non-bridging oxygens** (those linked to only one Si atom) per SiO_4 group. This number p varies from 4 in olivine (in which SiO_4 tetrahedra are linked only indirectly through other cations) to 0 in quartz. Between these limits, silicate minerals exhibit a wealth of structural diversity, exceeded in complexity only by the chemistry of carbon.

8.1.1 Monomer silicates

Pursuing the analogy with organic polymers, olivine (Figure 8.1) would be called a **monomer**, because the basic unit of the polymer occurs in it uncombined. Silicates built of isolated SiO_4 tetrahedra are commonly known as **orthosilicates**. They include other minerals like garnet (e.g. grossular, $Ca_3Al_2Si_3O_{12}$), zircon ($ZrSiO_4$) and topaz ($Al_2SiO_4F_2$). Notice that in each of these formulae the ratio of Si to O is $1:4$, a universal characteristic of orthosilicates. All of the oxygens in these structures are non-bridging, so they share the value $p = 4$ (Table 8.1).

8.1.2 Dimer silicates

The structures of a few minerals involve pairs of SiO_4 tetrahedra linked through a single bridging oxygen ($p = 3$). The formula of this dimeric group, consisting of two SiO_4 groups less one oxygen ($= Si_2O_7$), can be seen in melilite ($Ca_2MgSi_2O_7$, Figure 8.1). The commoner mineral epidote contains both single (SiO_4) and double (Si_2O_7) tetrahedral groups.

Table 8.1 Silicate polymers.

Structural type	p	Z:O	Example	Formula
Orthosilicate (monomer)	4	1:4	Forsterite (olivine)	$Mg_2[SiO_4]$
Dimer	3	1:3.5	Melilite	$Ca_2Mg[Si_2O_7]$
Ring	2	1:3	Beryl	$Be_3Al_2[Si_6O_{18}]$
Chain silicates				
Pyroxene	2	1:3	Diopside (pyroxene)	$CaMg[Si_2O_6]$
Amphibole	1.5	1:2.75	Tremolite (amphibole)	$Ca_2Mg_5[Si_8O_{22}](OH)_2$
Sheet silicate	1	1:2.5	Muscovite (mica)	$KAl_2[AlSi_3O_{10}](OH)_2$
Framework silicate	0	1:2	Orthoclase (feldspar)	$K[AlSi_3O_8]$

8.1.3 Chain silicates

The sharing of two oxygens by each SiO_4 group produces **chains** of tetrahedra of indefinite length that form the skeleton of one of the most important mineral groups, the pyroxenes. The chains are kinked rather than linear because alternate tetrahedra stick out in opposite directions (Figure 8.2). Each silicon atom possesses two non-bridging oxygens ($p = 2$) and shares two bridging ones, so that the composition of the whole chain can be written $(SiO_3)_n$, where n is the number of tetrahedra in the chain. All pyroxenes therefore have SiO_3 (or Si_2O_6) in their chemical formulae, as for example in diopside $CaMgSi_2O_6$. The chains can be stacked against each other in different ways, allowing pyroxenes to crystallize in both the orthorhombic and monoclinic systems (Box 8.4).

The chains define the crystallographic c-axis in pyroxenes. Parallel to this run several prismatic cleavage planes, such as the perfect $\{110\}$ cleavage responsible for the characteristic perpendicular cleavages in the basal section. These cleavages reflect the stronger cohesion within each chain compared with the strength of bonding between chains.

RING SILICATES
An obvious alternative to forming an infinite chain is to link the ends of the chain into a ring. The minerals beryl ($Be_3Al_2Si_6O_{18}$, Figure 8.1), cordierite ($Al_3Mg_2Si_5AlO_{18}$) and tourmaline are examples of ring silicates, in which the basic structural element is a ring of six SiO_4 tetrahedra. The simpler ring silicates have the p value (2) and Si:O ratio (1:3) of the pyroxenes (Table 8.1).

DOUBLE-CHAIN SILICATES
The amphibole structure can be regarded as a pyroxene chain in which alternate tetrahedra share an oxygen with one neighbouring chain. This produces a double chain or band (Figure 8.2), leading to a marked prismatic, sometimes even fibrous, habit. Owing to the wider double chains, the conspicuous prismatic cleavages intersect at about $55°$, compared to $90°$ for the pyroxenes.

The double chain can be regarded as a row of hexagonal rings, not present in pyroxenes. These accommodate additional anions, usually **hydroxyl** (OH^-) or fluoride (F^-). Owing to the presence of these volatile constituents – the main chemical distinction between pyroxenes and amphiboles – the amphiboles are unstable at high temperatures, decomposing into pyroxene and vapour.

Of every four tetrahedra in the amphibole structure, two share two oxygens and the other two share three oxygens ($p = 1.5$). In principle this leads to the formula Si_4O_{11} or Si_8O_{22}, as in tremolite ($Ca_2Mg_5Si_8O_{22}(OH)_2$). Some of the tetrahedral sites may contain Al in place of Si (the ionic radius of aluminium is just small enough to admit it to tetrahedral as well as octahedral sites – Table 8.2). This will of

Figure 8.2 Simplified silicate structures in (a) pyroxene, (b) amphibole and (c) sheet silicates, seen in plan and end-on.

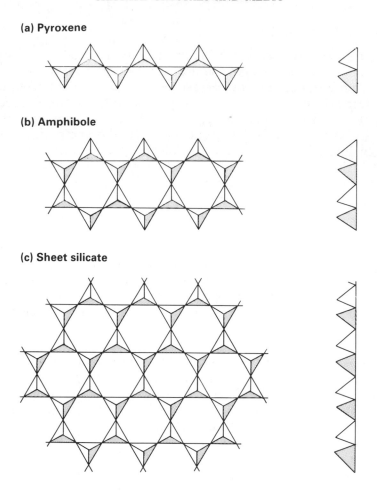

(a) Pyroxene

(b) Amphibole

(c) Sheet silicate

course alter the Si:O ratio in the amphibole formula, as in pargasite ($NaCa_2Mg_4Al[Al_2Si_6O_{22}](OH)_2$). Square brackets are used here to distinguish Al in tetrahedral sites from the remaining Al occupying octahedral sites. Notice that both of these examples conform to a more general formula Z_8O_{22}, where 'Z' includes both silicon and 'tetrahedral aluminium' (Box 8.4). Because of its different valency, the substitution of Al for Si requires an adjustment among other cations present (discussed later) in order to maintain electrical neutrality. A similar substitution occurs in the pyroxenes.

8.1.4 Sheet silicates

When every $(Si,Al)O_4$ tetrahedron shares three oxygens with neighbouring tetrahedra ($p = 1$), a continuous covalently bonded sheet structure is formed (Figure 8.2) which provides the basic framework for the micas – like biotite and muscovite – and a variety of other sheet

Table 8.2 Co-ordination in silicates.

Cation	Radius ratio	Predicted co-ordination	Occurrence in minerals
Si^{4+}	0.26	4-fold (tetrahedral) Z-site	
Al^{3+}	0.36		
	0.46		
Ti^{4+}	0.52		
Fe^{3+}	0.55	6-fold (octahedral) site	
Mn^{2+}	0.56		
Mg^{2+}	0.61		
Fe^{2+}	0.65		
Ca^{2+}	0.91	8-fold site	
Na^+	0.94		
K^+	1.27	$\geqslant$12-fold site	

Occurrence in minerals columns: OLIVINE, GARNET, PYROXENE, AMPHIBOLE, MICA, FELD-SPAR

silicates. All of them are 'hydrous', with OH^- or other anions occupying the rings, and like the amphiboles they 'dehydrate' a high temperatures (Figure 2.3).

Sheet silicates can be recognized chemically by having Z_4O_{10} in the formula (Table 8.1). The varieties of sheet silicate – chlorite, clay minerals, serpentine, etc. – can be regarded as multi-layer sandwiches, differing from each other in the identity of the ionic 'filling' and the manner in which sandwiches are stacked together. As in every sandwich, the bread is more coherent than the filling, so sheet silicates have a marked platy cleavage, as developed in the micas.

8.1.5 Framework silicates

Finally we come to a class of silicates in which every oxygen is shared between two tetrahedra, extending a semi-covalent network in all directions through the crystal. Because the volume of the crystal is determined wholly by the sp^3-based 'covalent' framework and not by the packing together of ions and separate SiO_4 polymers, the framework silicates usually have lower densities than other types of silicate (often less than $2.6 \, kg \, dm^{-3}$). The structures of some incorporate large cavities and

179

Box 8.2

Silicate melts

At the atomic level a silicate melt has much in common with silicate crystals: it consists of covalently bonded silicate structural anions such as $(SiO_4)^{4-}$ and $(Si_2O_7)_n^{6n-}$ glued together by cations like Mg^{2+} and Fe^{3+}. Over short distances there is the semblance of order (Box 1.2) but a silicate melt, like any liquid, lacks the **long-range order** characteristic of the crystalline state.

Another important difference is that, whereas nearly all silicate crystal structures are each constructed around a single type of silicate skeleton (chains in the case of pyroxenes, for example), the more open structure of a melt allows several types of silicate structural unit to coexist and intermingle. In **basic melts**, those rich in basic oxides and relatively poor in silica, a range of relatively simple polymers predominates $[SiO_4^{4-}, Si_2O_7^{6-}, Si_3O_{10}^{8-}, (SiO_3)_n^{2n-}$, etc]. This relatively unpolymerized structure favours the crystallization of minerals like olivine and pyroxene, because the appropriate polymers, at least in embryo form, are already present in the liquid. An **acidic melt** has a more polymerized structure conducive to the crystallization of minerals like mica (if water is present), feldspars and quartz.

The degree of polymerization of molten silicates is most clearly expressed in their viscosity. Viscosity increases with the degree of polymerization, because large polymeric units get tangled up with each other more than small ones, and this inhibits the flow of the liquid. Hence rhyolites (acid lavas) are much more viscous than basalts (Figure 3.8), even when allowance is made for temperature differences.

Dissolved water has a dramatic effect on magma viscosity, through its ability to sever polymers:

$$\begin{array}{ccccccc} | & | & | & & | & | & | \\ -O-Si-O-Si-O-Si-O- & + & H_2O & \rightarrow & -O-Si-O-Si-OH & + & HO-Si-O- \\ & & & & & & | \end{array}$$

Magma rich in dissolved water is therefore more fluid and conducive to diffusion than a dry magma of the corresponding composition. In this instance water is behaving as a basic oxide.

A high dissolved water content is only stable at high pressure, however. Close to the surface, ascending water-rich melt will become supersaturated and vesiculate to form bubbles of H_2O vapour. Viscosity then rises dramatically as the loss of water allows the magma to polymerize, and the presence of bubbles further enhances viscous behaviour (an extreme analogue being a stiffly whipped egg white).

---

channels through which cations and even molecules can diffuse quite readily.

As there are no non-bridging oxygens in the framework silicate structure ($p = 0$), the Z:O ratio is 1:2. The simplest composition is SiO_2, whose stable form at room temperature is quartz, specific gravity (SG 2.65); at elevated temperatures the more open structures of the polymorphs tridymite (SG 2.26) and cristobalite (SG 2.33) crystallize in its place. The three-dimensional network structure is reflected in the poor or non-existent cleavage of the silica minerals: quartz for example has a conchoidal fracture.

Substitution of Al into some of the tetrahedral sites in place of Si makes possible a huge variety of **aluminosilicate** minerals including the feldspars (the most abundant mineral group in the crust), and their silica-deficient cousins the feldspathoids. Replacing Si^{4+} with trivalent Al, of course, requires other ions to be introduced to maintain the charge balance. Owing to the openness of the framework structure, these compensating ions can be quite large (Na^+, K^+, Ba^{2+}).

Among the most open of the aluminosilicates are the zeolites. Unlike the feldspars they are hydrous, the water being held loosely in large intercommunicating cavities which can be up to 10 Å across. This framework is so rigid that the zeolites possess the remarkable ability to expel this water continuously and reversibly when heated, without their structure breaking down. Zeolites (natural and artificial) have many uses in chemical engineering, as ion exchangers and as molecular sieves that can separate small molecules according to their size. They perform a vital function as catalysts in the petroleum industry.

8.2   Cation sites in silicates

In most silicates (excluding the framework silicates) the tetrahedra fit together compactly, producing an orderly three-dimensional array of fairly closely-packed oxygen atoms. Si is known to occupy tetrahedral sites between these atoms. We can work out how the other ions such as Mg^{2+} and K^+ fit in by considering their radius ratios (Chapter 7). These are shown in Table 8.2, calculated from the ionic radii given in Box 7.2.

Two radii are given for Al^{3+} in Box 7.2 because, as the radius ratios suggest, its size permits it to fit into both tetrahedral and octahedral sites in silicates. (X-ray crystallography shows there is a slight difference in Al–O bond length in 4- and 6-fold co-ordination, so that a different Al^{3+} radius is associated with each site.) A glance at the formulae in Table 8.1 supports this conclusion. The formula of feldspar only makes sense if we include Al in Z, and X-ray investigations confirm that Al is tetrahedrally co-ordinated in feldspars. In beryl, on the other hand, it is octahedrally co-ordinated. In many minerals, as in mica, it is found in both sites.

The radius ratios of titanium (4+), iron (3+ and 2+), magnesium and manganese point to occupation of the octahedral ('Y') sites available in all **ferromagnesian** silicates. Ca^{2+} and Na^+ are somewhat larger, and require 8-fold co-ordination. Such ('X') sites do not exist in olivine, from which these elements are excluded (only traces of Ca are found), but they are available in pyroxene and amphibole (Box 8.4). K^+ clearly requires a still larger site. It occurs in amphiboles and micas in a large ('A') site associated with the silicate rings (Box 8.4) but this site does not exist in the pyroxene structure, which therefore excludes K.

In framework silicates there are no compact Y sites to accommodate Mg^{2+} and Fe^{2+}. Ca, Na and K occupy sites with somewhat irregular geometry, ranging in co-ordination number from 6 to 9. In the more

---

Box 8.3

Features of a silicate analysis

The table shows two ways of presenting the chemical analysis of a silicate (in this case an olivine).

Olivine analysis	Percentage of element*		Percentage of oxide†
Si	18.42	SiO_2	39.41
Fe (ferrous)	12.79	FeO	16.46
Mn	0.16	MnO	0.21
Mg	26.10	MgO	43.27
Ca	0.16	CaO	0.23
O	41.95		
Total‡			99.58

* The composition is expressed in percentages of each element (i.e. the number of grams of the element per 100 g of the sample). Such units are commonly referred to as 'weight percent', though 'mass percent' is a more accurate description. Oxygen appears as a separate item, but in fact it is unnecessary to analyse for it. The valency of each element requires that it combine with oxygen in **stoichiometric** proportions, even in complex silicates. It is possible to calculate the amount of oxygen present from the percentages of the other elements in the rock and their valencies.

† A more convenient format for including oxygen is simply to report the analysis in terms of the percentage of each oxide (g oxide per 100 g sample), as shown in the second column. There is no separate entry for oxygen, as it has all been allocated to individual oxides.

The entries in columns 1 and 2 are related to each other as follows:

$$\text{oxide}\,\% = \text{metal}\,\% \times \frac{\text{rel. mol. mass of oxide}}{n \times \text{rel. atomic mass}}$$

where n is the number of metal atoms per molecule of oxide. (It is necessary to specify FeO and Fe_2O_3 separately, or to assume that all of the iron is in one form or the other.)

‡ The oxide analysis is followed by a total, which for an accurate analysis should fall between 99.5 and 100.75% (all analytical measurements attract a small statistical error). If the total exceeds 100.75, one would suspect an unacceptable error in analytical procedure or calculation. A low total would suggest an error in the other direction, or neglect of an important constituent.

Only five oxides are significant enough to be listed in this analysis. In other minerals or rocks, a wider range of elements would require analysing. A typical silicate analysis would include the following **major elements** (those comprising more than 0.1% and included in the analysis total): SiO_2, Al_2O_3, Fe_2O_3, FeO, MnO, MgO, CaO, Na_2O. K_2O, TiO_2, P_2O_5, H_2O (see Exercise 2 at end of chapter) and CO_2. One might also look for certain **trace elements** (Ni being the prominent one in the case of olivine), whose concentrations would mostly fall below 0.1% and have no significant effect on the total (see Figure 9.1).

---

open structures of the zeolites these sites are larger still, and indeed may be substantially larger than the Na^+ and K^+ ions occupying them. This, in combination with the weak charge on these ions, means that they are readily removed and replaced by other ions, a process known as **ion exchange**.

8.2.1 Calculating site occupancies

The analysis of a silicate mineral is easier to understand if it is recalculated into a form directly comparable with the mineral's chemical formula. We have seen that in the formula of pure forsterite (Mg_2SiO_4), 2 magnesium ions and 1 silicon atom are associated with 4 atoms of oxygen. How many atoms of Mg (and Fe, Mn, Ca and Ni) and of Si would on average be associated with 4 oxygen atoms in a general olivine analysis like that given in Box 8.3? To answer this we need to recalculate the analysis in terms of the relative numbers of atoms, instead of mass percentages of oxides.

Table 8.3 shows how this calculation is carried out. The oxide analysis is written in column 1. The first step is to calculate the **number of moles** of each oxide present. This is achieved by dividing each oxide percentage by the appropriate relative molecular mass, entering the results in column 2. Because column 1 contains the number of grams of each oxide in 100 g of sample (i.e. mass percentage), column 2 contains the number of moles per 100 g.

Multiplying each entry in column 2 by the number of oxygen atoms in the corresponding oxide formula – 2 for SiO_2, 1 for the other oxides – gives the number of moles of O^{2-} associated with each oxide (column 3). Adding these up tells us that 100 g of sample contains a total of 2.6210 moles of O^{2-}. Our objective, however, is to calculate the numbers of cations associated with 4 moles of O^{2-}. Multiplying each entry in column 2 by $4/2.6210 = 1.5261$ gives us the number of moles of each oxide which together contain 4 moles of O^{2-} (column 4). As

Table 8.3 The mineral formula and site occupancies of an olivine.

	Relative molecular mass	1 Analysis as mass % oxides*	2 Analysis as moles of oxides[-]	3 Moles of oxygen (as O^{2+})[‡]	4 Cations per 4 oxygens[§]		
SiO_2	60.09	39.41	0.6558	1.3116	1.0008	Z site	1.001
FeO	71.85	16.46	0.2291	0.2291	0.3496 ⎫		
MnO	70.94	0.21	0.0030	0.0030	0.0046 ⎬ Y site	1.998
MgO	40.32	43.27	1.0732	1.0732	1.6378 ⎪		
CaO	56.08	0.23	0.0041	0.0041	0.0063 ⎭		
		99.58		2.6210			

$$\times \; \frac{4}{2.6210} \; =$$

* See Box 8.3.
[†] Column 1 divided by relative molecular mass.
[‡] Column 2 × number of oxygens per molecule (2 for SiO_2, 1 for the rest).
[§] Column 2 × 4/2.6210.

each oxide molecule contains only one atom of metal, the figures in column 4 also indicate the numbers of cations equivalent to a total of 4 oxygens.

The results in column 4 show two notable features. The first entry is a number very close to 1.0000. It represents the number of silicon atoms present for every four oxygen atoms in the olivine structure, and it indicates that the Z-sites in olivine are filled with silicon alone. Secondly the remaining entries in column 4 add up to 1.998. These elements collectively represent the average contents of the two Y-sites associated with every group of four oxygens in olivine. These conclusions allow us to write the complete chemical formula for the olivine, showing in what proportions the elements occupy each type of site:

$$(Mg_{1.638}Fe_{0.350}Ca_{0.006}Mn_{0.005})Si_{1.001}O_4$$

The close correspondence of the total site occupancies to the ideal formula of olivine (Y_2ZO_4) is additional reassurance of an accurate analysis.

Table 8.4 shows the formula calculation for an amphibole analysis.

Table 8.4 The mineral formula and site occupancies of an amphibole.

	Relative molecular mass	1 Analysis as weight% oxides	2 Analysis as moles* of oxides	2a Moles of metals[†]	3 Moles of oxygen (as O^2)[‡]	4 Cations per 24 oxygens[§]		
SiO_2	60.09	57.73	0.9607	0.9607	1.9214	7.786 ⎱ Z site		8.000
Al_2O_3	101.94	12.04	0.1181	0.2362	0.3543	IV 0.214 ⎰		
						VI 1.700 ⎱		
Fe_2O_3	159.70	1.16	0.0073	0.0146	0.0219	0.118		
FeO	71.85	5.41	0.0753	0.0753	0.0753	0.610 ⎬ Y site		5.056
MnO	70.94	0.10	0.0014	0.0014	0.0014	0.011		
MgO	40.32	13.02	0.3229	0.3229	0.3229	2.617 ⎰		
CaO	56.08	1.04	0.0185	0.0185	0.0185	0.150 ⎱ X site		1.975
Na_2O	61.98	6.98	0.1126	0.2252	0.1126	1.825 ⎰		
K_2O	94.20	0.68	0.0072	0.0144	0.0072	0.117	A site[¶]	0.117
H_2O	18.02	2.27	0.1260	0.2520	0.1260	2.042	OH site	2.042
Total		100.43			2.9615			

$$\times \frac{24}{2.9615} =$$

*Column 1 divided by relative molecular mass.
[†] Column 2 × number of cations in oxide molecule. (This column does not appear in Table 8.3 because there all oxide molecules had only one cation.)
[‡] Column 2 × number of oxygens per oxide molecule.
[§] Column 2a × 24/2.9615.
[¶] The A-site is only partly occupied (i.e. on the atomic scale, some A-sites are filled while others are empty.)

Box 8.4

Cation sites in pyroxenes and amphiboles

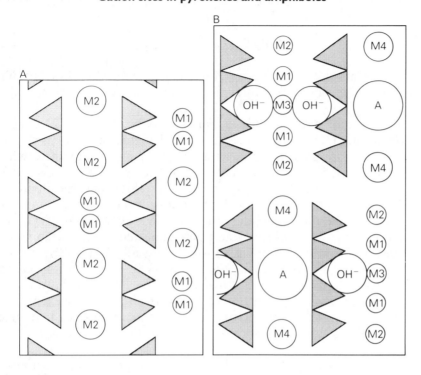

Diagram A shows a simplified end-on view of the pyroxene chains, indicating how they are stacked together, alternately back-to-back and point-to-point. In addition to the Z-sites (not shown) occupied by Si and a little Al, there are two types of cation site. Between two chains facing each other point-to-point lie two octahedral sites, designated M1, in which Al^{vi} ('octahedral Al'), Fe^{2+}, Fe^{3+}, Mg^{2+}, Mn^{2+}, Cr^{3+} and Ti^{4+} are accommodated. The other type of site lies between pairs of chains whose tetrahedra face each other base-to-base. They are called M2 sites and their geometry varies according to the ions occupying them. In the absence of Ca or Na, M2 is occupied by Mg^{2+}, Fe^{2+} and Mn^{2+} (for example in enstatite, $Mg_2Si_2O_6$), and has an irregular 6-fold co-ordination. The chains stack together in such a way as to produce an orthorhombic unit cell (orthopyroxene). The substitution of larger ions like Ca^{2+} or Na^+ causes a change of M2 geometry to 8-fold co-ordination. The presence of the larger ion disrupts the stacking, and forces the structure to become monoclinic, as in diopside ($CaMgSi_2O_6$).

Owing to the broader bands in the amphibole structure (diagram B), there are three slightly different types of octahedral site (two M1 sites, two M2 sites and one M3 site) in corresponding positions to the pyroxene M1 site. These are the five 'C' sites in the formula given in the main text. A larger site called M4 (or 'B' in the formula) corresponds almost exactly to M2 in the pyroxenes, accommodating small ions like Mg^{2+} in 6-fold co-ordination, and larger ions like Ca^{2+} and Na^+ in 8-fold co-ordination. The occupant of M4 plays the same rôle as M2 in the pyroxenes in determining the symmetry of stacking: Mg-rich amphiboles are commonly orthorhombic, whereas all other compositions are monoclinic. There are two M4 ('B') sites per formula unit.

The so-called A-site has no equivalent in pyroxenes. It lies sandwiched between pairs of bands whose undersides face each other, associated with the hexagonal rings which account for its large size. It is commonly unoccupied, but may contain Na^+ or K^+. The A-site lies opposite the OH-site, which also has no equivalent in the pyroxene structure.

The sites available in the amphibole structure (Box 8.4) are summarized by the formula:

$$A\ B_2\ C_5\ Z_8\ O_{22}\ (OH)_2$$

but in many amphiboles the large A site is partly or wholly vacant. The following points should be noted:

(a) The formula of an amphibole is normally written with 24 oxygens (including OH), so the analysis is recalculated on this basis.
(b) The ions are distributed between a greater variety of sites (which are discussed in Box 8.4). As a result the agreement between site occupancies and the ideal formula is less close than for olivine.
(c) There is insufficient Si to fill the 8 Z-sites-per formula unit. We assume that the remainder are occupied by Al ions (Al^{iv}), but most of the Al is left over and gets allocated to the octahedral C sites (Al^{vi}), in company with Fe^{3+}, Fe^{2+}, Mg^{2+} and Mn^{2+}.
(d) Ca^{2+} must be allocated to the larger B site, which also accommodates the Na^+.
(e) K^+ is too large to enter any but the A site. In this example, it falls a long way short of filling all of the A-sites, and in many amphiboles this site is vacant.

Formula calulations like these have several important applications in mineralogy. They help to confirm the accuracy of an analysis. Knowing what kind of crystallographic site an element occupies helps in understanding why, for example, pyroxenes contain no potassium, or why the structure of augite is different from hypersthene. Formula calculation also plays a part in classifying mineral groups that involve complicated solid solutions. The nomenclature of the amphiboles, for example, rests heavily on chemical parameters like the number of silicon atoms per 8 tetrahedral sites (item (c) above).

8.2.2 Effects of cation substitution

In the olivine solid solution series from the end-member forsterite (Fo, Mg_2SiO_4) to fayalite (Fa, Fe_2SiO_4), Mg ions in the sites are progressively replaced by Fe^{2+} ions. This **substitution** of Fe^{2+} for Mg^{2+} is possible because the ions have the same charge and similar size (Box 7.2). The larger radius of Fe^{2+} nevertheless causes a slight expansion of the crystal lattice: the z cell-dimension, for example, increases from 5.981 (Fo) to 6.105 Å (Fa). The substitution of Fe^{2+} for Mg^{2+} gives rise to similar, continuous solid solution series in garnets, pyroxenes, amphiboles and micas.

The effect is different when the substituting ion has a different size. Replacing Mg^{2+} in the pyroxene B-site with the larger Ca^{2+} ion forces the pyroxene to adopt a new structure of different symmetry. Because $Mg_2Si_2O_6$ and $CaMgSi_2O_6$ have intrinsically different structural arrangements, there cannot be a continuous series of compositions between them. There is a **miscibility gap** between these minerals except at high temperature, with the consequence that two pyroxenes of different composition can exist in the same rock in equilibrium with each other. A similar gap operates in the alkali feldspars between albite $(NaAlSi_3O_8)$ and orthoclase $(KAlSi_3O_8)$. The compositions of co-existing feldspars (or pyroxenes) lie on a **solvus** curve, as illustrated in Figure 2.6.

COUPLED SUBSTITUTION

A more complicated form of substitution is seen in plagioclase feldspar. The series of compositions from albite $(NaAlSi_3O_8)$ to anorthite $(CaAl_2Si_2O_8)$ reflects substitution in two sites at once. Substitution of Al^{3+} for Si^{4+} in tetrahedral sites leaves a charge imbalance, which is cancelled out by the **coupled substitution** of Ca^{2+} for Na^+ in the large cation sites. The overall substitution is of CaAl for NaSi. It follows that the Ca content of plagioclase correlates with the Al content, and the alkali (Na + K) content with Si. In the formula of any accurate plagioclase analysis (calculated to 32 oxygens), Si–(Na + K) will approximate to 8.0 (±0.1), and Al–Ca to 4.0 (±0.1). These requirements provide another measure of the quality of a plagioclase analysis (Exercise 3). Similar coupled substitutions occur in pyroxene – Na^+Al^{3+} for $Ca^{2+}Mg^{2+}$ in diopside to give jadeite, $NaAlSi_2O_6$, for example.

8.3   Optical properties of crystals

Crystal optics, a specialized subject for which several excellent textbooks are available, is mostly beyond the scope of this book. It is however relevant to take a brief look at how the optical properties of a crystal relate to its chemical bonding.

8.3.1   Refractive index

Light, like any electromagnetic signal, consists of synchronized oscillations of electric and magnetic fields (Boxes 5.2 and 6.3). It is the electric vector that is relevant to optical phenomena in crystals, because the progress of light in the crystal depends on the interaction between the electric disturbance and the atomic electron clouds. The extent of interaction, which is measured by the refractive index of the crystal, depends on the **polarizability** of the atomic assemblage. The highest refractive indices are found in minerals with metallic bonding, such as sulphides. Amongst the highest is galena, with a refractive index of 3.9.

8.3.2 Colour and absorption

Another important optical property is absorption. A mineral is **coloured** in transmitted light because certain wavelength components in the visible spectrum are more strongly absorbed than others, making the crystal appear to have the complementary colour (Figure 8.3). The region of absorption is generally a wavelength band rather than specific sharp lines. The green colour of olivine, for example, is due to absorption bands at each end of the visible spectrum, which emphasize the green wavelengths in the middle of the spectrum in the light emerging from the crystal.

The absorption of light energy is related to electron transitions between appropriately spaced energy levels in the crystal. The quantum energy of visible light is much smaller than that of X-rays (Chapter 6), and the transitions concerned occur between valence energy levels, either within an atom or between neighbouring ions. One very important factor in the colour of minerals is the splitting of d energy levels in transition-metal ions such as iron (Box 9.9). Elements like iron or manganese if present in a mineral play a major part in determining its colour. This can be seen in a solid solution series like phlogopite–biotite. The Mg-rich mica phlogopite ($KMg_3\{AlSi_3O_{10}\}\{OH\}_2$) is colourless or very pale in thin section, whereas the iron end-member biotite is intensely brown or red. Colour attributable to an essential constituent is said to be **idiochromatic**.

Certain other minerals, including some important gemstones, owe their colour to impurity elements. A common agent of such **allochromatic** colours is trivalent chromium, responsible for the deep red colour of ruby (a variety of corundum, Al_2O_3) and the green of emerald (a variety of beryl). The 'chrome diopside' which is conspicuous in many mantle-derived peridotites also owes its brilliant green colour to its Cr_2O_3 content (1% or so). Colour can also be due to the presence of

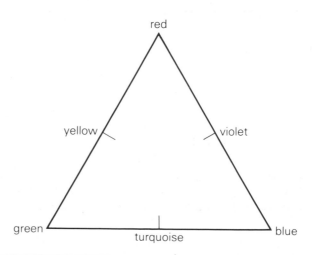

Figure 8.3 Colour triangle and complementary colours. Absorption of one colour will leave light relatively enhanced in the colour on the opposite side of the figure.

impurity phases. Quartz sometimes takes on a blue colour due to the presence of microscopic needles of rutile (TiO_2) or tourmaline.

Minerals may become coloured from being exposed to natural radiation for long periods. Such colours are due to radiation-induced defects in the crystal structure. The purple colour of fluorite, for example, is attributed to free electrons occupying vacant anion sites called 'colour centres'. An electron trapped in such a vacancy, like an atomic electron, is restricted to a number of quantized energy levels and transitions between these levels generate the observed colours. Heating eliminates most of the vacancies in a crystal and therefore bleaches the colour.

Metals and a number of important minerals, notably graphite and certain sulphides, absorb all light passing through them, so that in thin section they transmit no light at all. The property of **opacity** can be traced to metallic or quasi-metallic bonding in the crystal. Delocalized electrons in a metal behave like a charged fluid (Chapter 7). The alternating polarization of this electron fluid, induced by the incoming light, involves work being done in moving electrons around, and this dissipates the energy of the incident light (transforming it into thermal motion) just as the energy of a walker is sapped by walking across unconsolidated dry sand. A seismological analogy is the absorption of shear waves by liquids, which is why such waves do not pass through the Earth's core whereas compressional waves do. Insulating materials like quartz, in which there are no delocalized electrons, are electromagnetically elastic: electrons fixed in atoms cannot be moved around and absorb practically no energy from the beam, which is therefore transmitted with little loss of intensity like seismic waves through solid rock.

8.3.3 Reflectivity

The striking characteristic of minerals like pyrite and galena is their metal-like **reflectivity**. A reflected-light microscope with a special reflectivity accessory can be used to measure quantitatively the proportion of the perpendicularly incident light that is reflected back at various wavelengths, and this provides a vital diagnostic tool for the ore mineralogist. The reflectivity of a mineral increases with its refractive index and with its opacity (Bloss, 1971, Figure 12.13).

8.3.4 Anisotropy

Whereas minerals like halite and diamond have optical properties which are uniform in all directions, the majority of minerals are optically **anisotropic**. That is to say, the refractive index, the colour and – for ore minerals – the reflectivity vary according to the direction in which the incoming light (specifically its electric vector) oscillates.

Incoming light entering a birefringent mineral is split into two rays with mutually perpendicular vibration directions, which experience different refractive indices as they travel through the crystal. The **birefringence** of the crystal is the difference between its maximum and minimum refractive indices. The mineral calcite is among the most birefringent of the common minerals, to the extent that the phenomenon can even be seen without the aid of a microscope. The carbonate anion CO_3^{2-} has a symmetrical planar shape (Figure 8.4), reflecting the geometry of the sp^2-hybrid that the carbon atom uses to form σ-bonds with the three oxygen atoms. All the CO_3^{2-} ions in calcite have the same orientation perpendicular to the three-fold symmetry (z) axis of the crystal. The remaining carbon p-electron can establish a π-bond with any one of the three oxygens, giving three possible configurations. As with graphite (Box 7.4), the real configuration is a blend of all three. Y-shaped π molecular orbitals therefore exist above and below the plane containing the nuclei. To an electric field oscillating perpendicular to the z-axis [Figure 8.4(a)], the anion looks highly polarizable, because delocalized electron density is easily shunted from one end of these π-orbitals to the other by the oscillating electric field. Light having this electric vibration direction therefore experiences a relatively high refractive index (1.66). The polarizability parallel to the z-axis is much lower, so light vibrating in this direction encounters a lower refractive

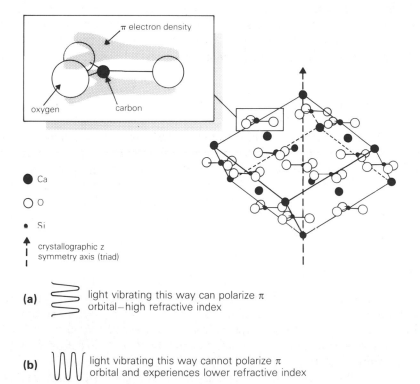

Figure 8.4 Structure of calcite showing (inset) the orientation of sp^2-based trigonal planar CO_3^{2-} oxy-anion. The highly polarizable concentrations of π electron density perpendicular to the z-axis are shown stippled. Light vibrating in plane (a) encounters a high refractive index. Light vibrating parallel to the z-axis (b) causes less polarization and experiences a lower refractive index.

π electron density

oxygen

carbon

● Ca

○ O

• Si

↑ crystallographic z symmetry axis (triad)

(a) light vibrating this way can polarize π orbital – high refractive index

(b) light vibrating this way cannot polarize π orbital and experiences lower refractive index

index (1.49). Therefore when a mark on a piece of paper is viewed through a clear cleavage rhomb of calcite, two separate images are seen.

The majority of silicate minerals are also optically anisotropic for similar reasons, but the degree of anisotropy (the magnitude of the birefringence) is much lower than for carbonates.

8.4 Defects in crystals

The extreme order of the crystalline state conceals the fact that all crystals incorporate structural defects that have a profound effect on the growth of crystals and on their mechanical strength.

8.4.1 Crystal growth

The first step in the production of a crystal from a surrounding liquid (melt or solution) is **nucleation**, the accretion of a minute ordered nucleus upon which the rest of the crystal will be deposited. The free energy of such a nucleus consists of

(a) a negative term proportional to the volume, reflecting the cohesive forces between close-packed ions/atoms in the interior; and
(b) a positive term proportional to the surface area, reflecting the reactivity of unsatisfied bonding potential on the surface.

Thus for a cubic nucleus of edge-length r:

$$\Delta G_{\mathrm{L}} = -r^3 L_{\mathrm{v}} + 6r^2 \sigma_{\mathrm{s}}$$

where ΔG_{L} is the free energy of the nucleus relative to an equivalent amount of melt. L_{v} is the free energy of fusion per unit volume and σ_{s} is the surface energy per unit surface area (Figure 8.5).

The initial nucleus, having a high surface area:volume ratio, is therefore highly unstable, and it must grow rapidly into a larger, more stable crystal if it is not to be redissolved by the liquid. In practice, therefore, melts cool below the liquidus, becoming supersaturated with the crystal species, before the first visible crystals appear, a phenomenon called **supercooling**.

The subsequent growth of a crystal can be seen in terms of adding new material (depicted as rectangular blocks) in layers on an existing crystal surface, as depicted in Figure 8.6a. The blocks can represent ions, molecules or entire unit cells. Exposed corners and edges are features of very high surface energy owing to the number of unsatisfied bonds. Adding a block at A is therefore highly unfavourable in energy terms, and would occur only in a strongly supersaturated solution. Sites B and C are less hostile, but continued exploitation of such step sites will complete the layer and eliminate the step. In subsequent cryst-

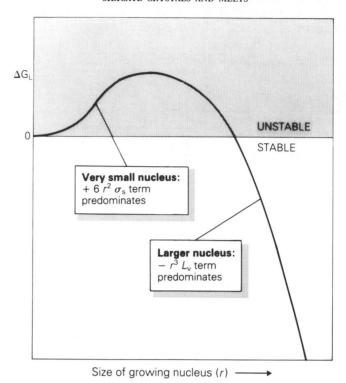

Figure 8.5 Energetics of crystal nucleation. ΔG_L is the free energy of the nucleus relative to the equivalent amount of melt.

SILICATE CRYSTALS AND MELTS

ΔG_L

0

UNSTABLE

STABLE

Very small nucleus: $+ 6\,r^2\,\sigma_s$ term predominates

Larger nucleus: $- r^3\,L_v$ term predominates

Size of growing nucleus (r) ⟶

allization on this face, use of sites like A is unavoidable. Calculations suggest that a high threshold of supersaturation or supercooling has to be surmounted before crystallization at sites like A can proceed and crystallization ought therefore to be an extremely slow process, yet in practice this restriction does not seem to operate.

The explanation of the discrepancy lies in a lattice imperfection called a **screw dislocation** (Figure 8.6b). Such defects are stacking 'mistakes' incorporated randomly into crystals and perpetuated during growth. A mismatch of layers on one side of the crystal is not seen on the other side, and must disappear at a line (dislocation) extending vertically through the middle of the crystal, where one layer actually twists up into the next one. The crystal resembles the kind of multi-storey car park in which the floors are part of a large spiral leading to the top. The vital feature is that the step affording favourable crystallization sites like B and C is perpetuated as crystallization proceeds, spiralling continuously upward as suggested by the arrows and the crystallization 'fronts' in Figure 8.6c. Such dislocations make continued crystallization practicable at modest degrees of supersaturation. Specialized photographic techniques can indeed detect spiral growth patterns on artificially grown crystals (Fyfe, 1964, Figure 13.8).

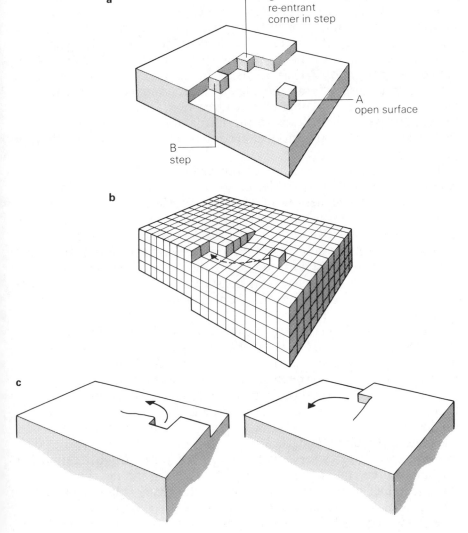

Figure 8.6   a. Sites of accretion on a crystal face. b. Screw dislocation. (Reproduced with permission from W. T. Read, Jr (1953) *Dislocations in crystals.* Copyright McGraw-Hill.) c. Subsequent positions of step as crystallization proceeds.

a

C
re-entrant
corner in step

A
open surface

B
step

b

c

8.4.2   *Mechanical strength of crystals*

Box 7.7 includes estimates of the strength of bonding in several crystalline materials. From data such as these it is not difficult to calculate values for the shear strength of a number of structurally simple crystalline materials, such as pure metals. However carefully such calculations are carried out, the results are generally 100–1000 times greater than the measured shear strength for the material in question. In resolving this embarrassing discrepancy, we have to recognize the rôle of another type of crystal imperfection, the **edge dislocation**.

An edge dislocation (Figure 8.7a) marks the edge of an extra half-layer in a crystal lattice. This may represent a layer overstepped during growth of the crystal, or may be the result of deformation. Note that bonds in the immediate vicinity of the dislocation are either stretched or compressed, and owing to this departure from equilibrium length are not as strong (Box 7.1) as bonds in the undistorted lattice. The high free energy associated with the edge dislocation makes it a particularly susceptible site for initiating chemical reactions: acid etching of freshly broken crystal surfaces leads to the formation of pits at points where dislocations intercept the surface, making them visible under an electron microscope.

The mechanical significance of the dislocation becomes clear if we consider the crystal being subjected to **shear stress** as shown in Figure 8.7b. The crystal at first responds with **elastic strain**, in which it is deformed to a minute and reversible extent by the stretching and compression of bonds, without any being broken. Permanent shear deformation (**ductile deformation**) requires bonds to be broken so that the top half of the crystal can be bodily moved to a new position across the lower half. A perfect crystal resists this very effectively, because a whole layer of bonds must be broken simultaneously, requiring a colossal input of energy.

The presence of an edge dislocation changes this picture dramatically. When the shear stress is applied, the dashed row of bonds immediately to the right of the dislocation (already stretched in the unstressed crystal) becomes more stretched than other bonds, and will be the first to rupture. This will happen when the distance from row a to row b becomes less than the stretched bond length $b-c$. These bonds will at this point flip to link rows a and b instead. Row a is now part of a complete layer, and the dislocation has migrated to row c. If the stress is maintained, this process will be repeated until the dislocation has migrated to the edge of the crystal. The result is the net movement of the upper half of the crystal over the bottom half (Figure 8.7b), in a manner that requires *only one row of bonds to be broken* at one time. The stress required is orders of magnitude less than that needed to deform the perfect crystal. Edge dislocations therefore explain why the shear strength (and tensile strength) of crystalline materials is less than theoretically expected.

A square centimetre of any crystal intersects 10^8-10^{12} edge dislocations. The exact number depends on the crystal's deformation history. Deformation – 'working' in the metallurgist's jargon – causes dislocations to multiply and initially makes the crystal more ductile.

Most minerals are considerably more brittle than metals. At high temperatures and pressures, however, **dislocation creep** can become an important mechanism of rock and mineral deformation. Its effect is to make silicate rocks behave in a ductile (plastic) manner if subjected to stress over long periods of time. It thereby becomes possible for solid mantle rocks to **convect** (migrate in response to temperature-induced

Figure 8.7 a. Edge dislocation. (Reproduced from Bloss (1971), with permission of Holt, Rinehart and Winston, Inc.). b. Migration of edge dislocation in response to shear stress (arrows). The stippled portion of the upper layer **slips** over the lower layer (s–s is the **slip plane**) as the edge dislocation (heavy stipple) jumps from row to row.

density gradients), the phenomenon upon which terrestrial heart flow and plate tectonics largely depend. If all crystals were perfect, the Earth would be a very different planet.

8.5 Bibliography

Bloss, F. D. (1971) *Crystallography and crystal chemistry*, Holt, Rinehart and Winston New York.

Cox, K. G., Price, N. B. and Harte B. (1988) *The practical study of crystals, minerals and rocks* revised edn. McGraw-Hill, London.

Evans, R. C. (1964) *An introduction to crystal chemistry*, Cambridge University Press, Cambridge.

Fyfe, W. S. (1964) *Geochemistry of solids*, McGraw-Hill, New York.

Klein, C. and Hurlburt, C. S. Jr (1993) *Manual of mineralogy* (21st edn after J. D. Dana), Wiley, New York.

8.6 Exercises

1 Predict the degree of Si–O polymerization of the following minerals:

Edenite	$NaCa_2Mg_5(AlSi_7O_{22})(OH)_2$
Hedenbergite	$CaFeSi_2O_6$
Paragonite	$NaAl_2(AlSi_3O_{10})(OH)_2$
Leucite	$K(AlSi_2O_6)$
Acmite	$NaFeSi_2O_6$

2 Express the first analysis in Question 3 in terms of element percentages.

3 Calculate the site occupancies of the following mineral analyses, matching your results with the ideal formulae given (cf. Tables 8.3 and 8.4). Carry out all calculations to 4 decimal places.

	Garnet* $X_3Y_2Z_3O_{12}$	Epidote[‡§] $X_3Y_2Z_3O_{12}(OH)$	Pyroxene[¶] $M_2Z_2O_6$	Feldspar[‖] $X_4Z_8O_{32}$
SiO_2	38.49	39.28	46.92	52.73
Al_2O_3	18.07	31.12	3.49	29.72
TiO_2	0.55[†]	–	1.19	–
Fe_2O_3	5.67	4.15	0.95	0.84
FeO	3.76	0.42	20.31	–
MnO	0.64	0.01	1.13	–
MgO	0.76	0.01	7.30	–
CaO	31.59	23.44	17.35	12.23
Na_2O	–	–	1.24	4.19
K_2O	–	–	–	0.13
H_2O^+	–	1.87[§]	–	–
Total	99.63	100.30	99.84	99.84

* In the garnet structure, all divalent ions reside in the 8-fold ('X') site.
[†] Rel. mol. mass: 79.9. (Others given in Table 8.4.)
[‡] Calculate to 13 oxygens (12 + one OH group).
[§] The *total* water content of an analysis consists of:
    (a)   Internal structural 'water' (actually present as OH but measured as H_2O). This is generally assumed to be released only at temperatures above 110 °C. Denoted H_2O^+.
    (b)   Adsorbed water (moisture adhering to the surface of powder grains). Released below 110 °C ('H_2O^-').
[¶] Fill the Z site with sufficient Al to make 2.0000. The remaining Al goes in M. In a pyroxene formula it is difficult to distinguish between M1 and M2 (Box 8.5), because Mg, Fe^{2+} and Mn enter both sites. Here they are considered together as two 'M' sites.
[‖] Si, Al and Fe^{3+} reside in the Z-site.

4 Plot the pyroxene composition (Exercise 3) in a ternary diagram whose vertices represent Ca, Mg and $(Fe^{2+} + Fe^{3+} + Mn)$. The method is described in Appendix B (use commercial triangular graph paper or draw your own 20 cm equilateral triangle on an A4 sheet). Plot the ideal compositions of diopside, hedenbergite, enstatite and ferrosilite $(Fe_2Si_2O_6)$.

9

Some geologically
important elements

Chemical elements are of interest to the geologist for a variety of reasons. Some, such as silicon (Si) and iron (Fe), are so abundant that their chemical properties govern the behaviour of geological materials. Less abundant elements like rubidium (Rb) and strontium (Sr) participate more passively in geological processes, yet may help us to understand how such processes work. Other elements (such as chromium, Cr, or uranium, U) have important commercial uses and therefore attract the attention of the exploration geologist.

The purpose of this chapter is to introduce the reader to the geochemistry of the more important chemical elements, to show how the Periodic Table can be used to predict element behaviour in the Earth and on its surface as well as in the laboratory, and to illustrate what these elements and their isotope systems can tell us about Earth processes. In view of the tendency for more geologists to find employment in the environmental industry, the emphasis here will be as much environmental as geological.

9.1 Major and trace elements

In terms of their abundance in geological materials, elements can be divided into two classes: **major elements** and **trace elements**.

MAJOR ELEMENTS

These elements, which include Si, Al, Mg and Na, have concentrations in most geological materials in excess of 0.1%. They are essential constituents of rock-forming minerals. Major element concentrations in silicate minerals are usually expressed in terms of **oxide percentages** (Box 8.3). The term **minor element** is sometimes applied to the less abundant major elements, such as managanese (Mn) and phosphorus (P), with oxide concentrations below 1%.

TRACE ELEMENTS

Trace elements, such as rubidium (Rb) and zinc (Zn), have concentrations in most geological materials too low – usually less than 0.1% – for them to influence which minerals crystallize. They may occur as dissolved 'impurities' in rock-forming minerals, or in separate accessory minerals. Their concentrations in rocks are usually expressed in **parts per million** (ppm) of the element (not oxide) or parts per **billion** (Figure 9.1).

This distinction is applied in a flexible way. The same element can be a major element in one rock type (potassium in granite, for example) and a trace element in another (potassium in peridotite).

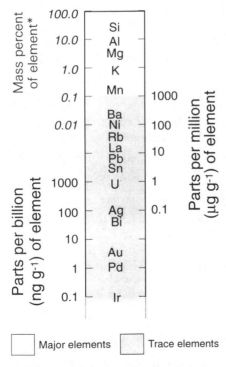

Major elements   Trace elements

* Major elements usually expressed as mass percent oxide (Box 8.3).

Figure 9.1 How units of element concentration compare. Element symbols illustrate the average concentration of selected elements in continental crust.

9.2   Alkali metals

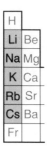

The alkali metals comprise the first column (Group 1a) on the left-hand side of the Periodic Table. Their key chemical features are as follows:

(a)   they are strongly electropositive elements; their compounds are characteristically ionic and they form basic oxides.

(b)   the M^+ cations are large and can be accommodated only in relatively large cation sites, such as the 'A-site' in amphiboles and micas (Chapter 8) and in tektosilicate minerals. Feldspar is the main host of these elements in most rocks.

(c)   they are very soluble in aqueous fluids, and are among the first elements to be dissolved during weathering. Na and K are important constituents of sea water, and evaporite deposits provide the main industrial source.

(d)   potassium is an important plant nutrient, hence its widespread use in fertilizers.

Sodium (Na) and potassium (K) are important constituents of feldspar, amphibole and mica and therefore are found as major elements in most rocks of the continental crust. Rubidium (Rb) and cesium (Cs), on the other hand, are too scarce to form their own minerals and occur only as trace elements, entering rock-forming silicates only where they can substitute for K^+ ions (e.g. in alkali feldspar).

The large ionic radii of K^+, Rb^+, Cs^+ and to a lesser extent Na^+ lead to their exclusion from dense ferromagnesian minerals like olivine and pyroxene. Such minerals, with calcic plagioclase, are the first to crystallize from basic magmas, and because the alkali metals are excluded they remain in the melt. As the amount of melt decreases with advancing crystallization, the concentration of dissolved K, Rb and Cs in the remaining melt increases. A series of lava flows tapping a magma chamber at successive stages in its crystallization would therefore exhibit increasing concentrations of these elements with time. Among plutonic rocks they are enriched in 'late-stage' granites and pegmatites. Elements like these, whose exclusion from the main igneous minerals leads to their concentration in late-stage residual magmas, are called **incompatible elements** (Box 9.1).

9.2.1   Radioactive alkali metals

Two of the alkali metals have radioactive isotopes of great significance to the geologist: ^{40}K and ^{87}Rb. Both have long half-lives (1.25×10^9 and 48.8×10^9 years respectively), roughly comparable to the age of

Box 9.1

Incompatible elements

Incompatible elements are those whose ions are not easily accommodated in the structures of the principal igneous or metamorphic minerals. During magmatic crystallization they become progressively more enriched in the diminishing amount of residual liquid, whose disordered structure accepts them more readily.

Major elements are shown **bold.**

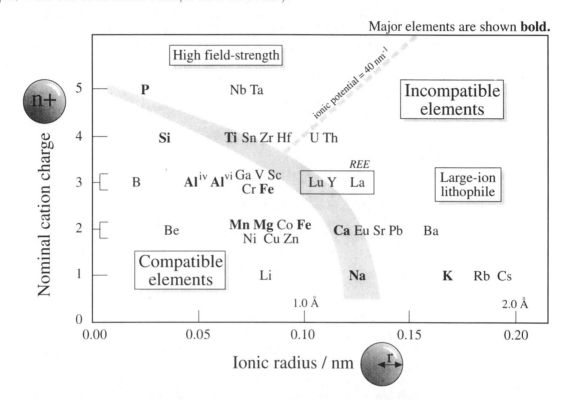

The stippled band divides elements whose ions are readily accommodated in relevant igneous minerals ('compatible elements') from elements that are predominantly incompatible. Elements whose symbols overlap the band may fall into either group depending on the minerals crystallizing.

The reason why an incompatible element is excluded from the crystal structure depends on its ionic potential (Figure 7.7):

Large-ion lithophile elements
(Ionic potential $<40\,nm^{-1}$.) These are elements like Rb and Ba, which most crystals exclude because their ions are simply too large to fit into the sites available (Box 7.2).

High field-strength elements
(Ionic potential $\geqslant 40\,nm^{-1}$.) An ion like zirconium (Zr^{4+}) has a radius no bigger than Mg^{2+}, yet its high polarizing power and relatively covalent bonding make it an uncomfortable occupant of a Mg^{2+} cation site in a predominantly ionic crystal.

The behaviour of the **rare earth elements** (REE = La to Lu) and yttrium (Y) is discussed later (p. 229).

the Earth (4.6×10^9 years). The amounts of these isotopes in the Earth are decreasing with time, but there has been too little time since the last episode of element-formation (Chapter 10) for them to have decayed away completely. The rates of decay are accurately known from laboratory measurement, and ^{40}K and ^{87}Rb provide isotopic clocks that can be used for dating gelogical events (Box 9.2; Faure, 1986).

Radioactive decay generates heat (Box 10.2). Much of the heat currently escaping from the Earth's interior is due to the decay of the radioactive isotopes of potassium (K), thorium (Th) and uranium (U) in the crust and mantle (Box 10.1). ^{40}K is thought to be responsible for about 15% of the heat generated in the crust. Owing to its low abundance and slow rate of decay (long half-life), ^{87}Rb does not make a significant contribution to the Earth's heat flow.

9.3 Hydrogen

The position of **hydrogen** (H) in Group I of the Periodic Table suggests that it should behave like the alkali metals, but in fact the similarity is restricted to valency. The ionization energy of hydrogen is much higher than the alkali metals (Figure 6.1b), and because the valence shell consists of the 1s orbital alone hydrogen possesses no metallic properties (except under extreme conditions, such as within the interior of Jupiter – p. 162).

The predominant hydrogen compound on Earth is the oxide H_2O, which occurs in the familiar gas (vapour), liquid and solid forms (Box 2.2, figure B). The presence of liquid-water oceans on the Earth's surface is unique in the Solar System, and has been essential to the evolution of life on Earth. The peculiarities of the ice–water system play an important part in regulating the Earth's climate (Box 4.1) and in driving erosion and the transport of sediment.

Within the Earth, hydrogen occurs in rocks, mainly in the form of OH^- ions in hydrous minerals such as muscovite [$KAl_3Si_3O_{10}(OH)_2$]. Heating such minerals leads to dehydration (Figure 2.3), releasing H_2O-rich fluids that are essential agents in regional metamorphism. The presence of a hydrous fluid also lowers the temperature (the solidus) at which a rock begins to melt, and such fluids therefore play an important role in bringing about melting in the mantle and crust above subduction zones; the water originates by dehydration of hydrous minerals in the altered oceanic crust of the downgoing slab.

Water, as well as accelerating many geochemical reactions (Chapter 3), exerts a powerful influence on the physical properties of melts (Box 8.2). The explosive expansion of steam escaping from ascending

Box 9.2

Radiometric dating: K–Ar geochronometry

Present isotopic composition of potassium:

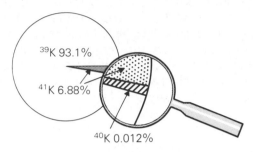

^{39}K 93.1%

^{41}K 6.88%

^{40}K 0.012%

^{40}K is radioactive.

Some ^{40}K nuclei decay to ^{40}Ar and others to ^{40}Ca:

$$^{40}K \xrightarrow{\text{decay}} \begin{array}{l} ^{40}\text{Ar} \quad \lambda_{Ar} = 0.585 \times 10^{-10} \text{ yr}^{-1} \\ ^{40}\text{Ca} \quad \lambda_{Ca} = 4.72 \times 10^{-10} \text{ yr}^{-1} \end{array}$$

$$[\lambda = \lambda_{Ar} + \lambda_{Ca}]$$

Crystal of
potassium mineral

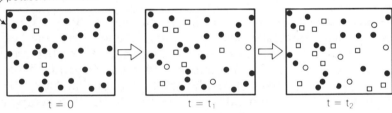

t = 0   t = t$_1$ t = t$_2$

● ^{40}K nuclei
○ ^{40}Ar nuclei
□ ^{40}Ca nuclei

Atoms of ^{40}K
scattered through
a new crystal
of a K mineral.
A few ^{40}Ca atoms
are also present.

As time passes, ^{40}K nuclei decay into ^{40}Ca or ^{40}Ar nuclei. The ratio ^{40}Ar : ^{40}K increases with the time that has elapsed since t = 0. If no ^{40}Ar escapes, the ratio can be used to calculate the age of the crystal:

$$t = \frac{1}{\lambda} \ln \left[1 + \frac{\lambda}{\lambda_{Ar}} \frac{^{40}\text{Ar}}{^{40}\text{K}} \right]$$

Daughter/parent
ratio today

magma, as it experiences depressurization close to the surface, provides the energy for the most destructive volcanic eruptions on Earth.

Natural hydrogen consists of two stable isotopes ^1H (99.984%) and ^2H (0.016%). The latter, known by the name **deuterium** (the chemical symbol D is synonymous with ^2H), has 1 proton and 1 neutron in the nucleus. The abundance ratio of these two isotopes varies slightly in different types of natural water, and serves as a pointer to the origin of solutions involved in geological processes (Box 9.7). Hydrogen has a third isotope, **tritium** (^3H: 1 proton + 2 neutrons), which is radioactive with a half-life of about 12 years.

9.4   Alkaline earth metals

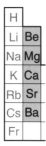

Relations between the divalent alkaline earth elements bear some resemblance to those between the alkali metals. **Beryllium** (Be), like lithium, has rather different chemical properties from the other members of the group (Box 9.3). **Magnesium** (Mg) and **calcium** (Ca) are major elements in most rock types, whereas **strontium** (Sr) and **barium** (Ba) are trace elements showing degrees of incompatible behaviour. All are strongly electropositive, reactive metals. The alkaline earths form stable, highly refractory, basic oxides.

The ionic radius of Mg^{2+} is similar to that of Fe^{2+} (Box 7.2). Ferromagnesian minerals such as olivine and pyroxene exhibit complete solid solution between a magnesian end-member (such as forsterite, Mg_2SiO_4) and a ferrous end-member (fayalite, Fe_2SiO_4). Because MgO is a refractory oxide (melting point 2800 °C), the Mg end-member of these minerals has the higher melting temperature (e.g. Box 2.4). The crystallization of ferromagnesian minerals from a magma depletes it in MgO more rapidly than FeO, so the FeO : MgO ratio of the residual melt increases with advancing crystallization and provides a useful indication of the degree of fractionation of a magma.

Magnesium is an essential constituent of chlorophyll (Box 9.6), and therefore plays a part in photosynthesis. After Na^+, Mg^{2+} is the most abundant cation in sea water (Table 4.3) and is recovered from it industrially. Though Ca^{2+} is less abundant in sea water, Ca salts (carbonate and sulphate) are among the first to precipitate on evaporation; few evaporite sequences include magnesium minerals, which appear at a much more advanced stage of evaporation. Most of the calcium in sediments, however, occurs in biogenic limestone.

Box 9.3

Lithium, beryllium and boron

Lithium, beryllium and boron are chemically somewhat different from the other members of their respective groups. Except for the difference in valency, beryllium has more in common chemically with aluminium than with magnesium or calcium. The reason for this diagonal relationship is that beryllium's ionic potential is much closer to that of aluminium than to the other alkaline earth metals (see diagram), leading to similar, relatively covalent bonding behaviour and similar crystal chemistry. Both Be and Al form unreactive, very hard oxides that are chemically amphoteric (Box 8.1). Both metals resist acid attack and oxidation – unlike Mg, for example – owing to the formation of a surface film of oxide.

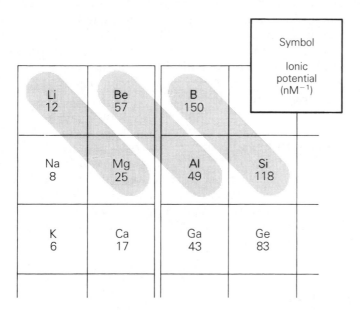

For the same reason boron has much in common with silicon. Both are semiconducting metalloids rather than true metals. B_2O_3 is an acidic oxide like SiO_2. The diagonal relationship between lithium and magnesium, though less pronounced, means that Li is found mainly in the Mg sites in silicate minerals.

Li, Be and B also differ from other elements in their groups in having anomalously low abundance, in the Earth and in the solar system (Chapter 10). For instance, Li is a trace element in most rocks, whereas Na and K are major elements.

9.5 Aluminium

Aluminium is the commonest metallic element in the crust (Figure 10.7). Its malleability, ductility, electrical conductivity, stability and low density make it ideal for many industrial and domestic uses. The oxide, alumina (Al_2O_3), is amphoteric. It forms the extremely hard, refractory mineral **corundum** (hardness = 9), which is widely used as an abrasive (emery) and in ceramics. The gemstones ruby and sapphire are coloured varieties of corundum.

The mobility of aluminium in weathering depends on the pH of the solution. Al_2O_3 is extremely insoluble in the majority of ground waters, whose pHs tend to lie in the range 5–6 (Figure 4.1b). During the weathering of a granite under such conditions, Na and Ca can be almost entirely removed and K, Mg, Fe^{3+} and even Si partially leached, while Al remains immobile. The minerals left behind are therefore very aluminous, as in the china clay deposits (consisting largely of kaolinite) associated with Cornish granites. The weathering of intermediate volcanics and ash in extreme tropical conditions can proceed even further and remove most of the silica too, leaving behind a mixture of aluminium hydroxides stained by hydrated iron oxides. This material, the chief industrial source aluminium, is called **bauxite**.

Contact with rotting vegetation or pollution by acid rain can, however, make surface waters sufficiently acid (pH 4 or less) to dissolve alumina. Signs of this can be seen in soil profiles in temperate forests, where an upper light-coloured horizon, from which all components except silica have been leached by acid solutions, passes down into a lower horizon rich in clay minerals, in which Al has been re-precipitated through contact with relatively neutral ground water at the water table. Precipitation of Al^{3+} (and Fe^{3+}) in this way is an example of **hydrolysis** (Box 9.4).

9.6 Carbon

Carbon is unique among the chemical elements for the huge variety of compounds it can form: the number of known carbon compounds exceeds the number of compounds associated with all of the other elements considered together. So extensive is the chemistry of carbon and so vital is it to life that, on its own, it constitutes a separate branch of chemical science called **organic chemistry**. Sedimentary rocks commonly contain 0.2 to 2% of organic matter, the highest concentrations occurring in shales.

From the point of view of its occurrence in rocks, it is helpful to consider the chemistry of carbon under two headings: organic carbon and inorganic carbon.

Box 9.4

Hydrolysis

To a geochemist, the term **hydrolysis** refers to reactions in which either (or both) of the O–H bonds in water is broken. Consider the hydrolysis of atmopheric sulphur dioxide.

$$SO_2 + H_2O \rightarrow H_2SO_3 \rightarrow H^+ + HSO_3^-$$
$$\text{sulphurous acid}$$

This is one of the reactions contributing to the phenomenon of **acid rain** (O'Neill, 1992). Hydrolysis of an acidic oxide (SO_2, NO_2, CO_2) generally gives an acidic solution. The converse applies to the hydrolysis of a basic oxide:

$$Na_2O + H_2O \rightarrow \qquad 2NaOH$$
$$\text{basic hydroxide solution}$$

Hydrolysis reactions are important in weathering. Elements of intermediate ionic potential like Al and Fe(III) are soluble only in fairly acid solutions. If an Al-bearing solution mixes with a less acidic solution, precipitation occurs as a result of hydrolysis:

$$Al^{3+} + 3H_2O \rightarrow Al(OH)_3 + 3H^+$$
$$\text{solution} \qquad\qquad \text{insoluble}$$
$$\text{precipitate}$$

Geochemists refer to such elements and their precipitates as **hydrolysates** (see diagram). For certain elements like Fe and Mn, the first step in this process is oxidation. Fe^{2+} and Mn^{2+} are readily dissolved by mildly acid waters during weathering, but once dissolved they are prone to oxidation to Fe^{3+} and Mn^{4+}, whose higher ionic potentials lie in the hydrolysate field and which consequently precipiate, as minerals like goethite (FeO(OH)).

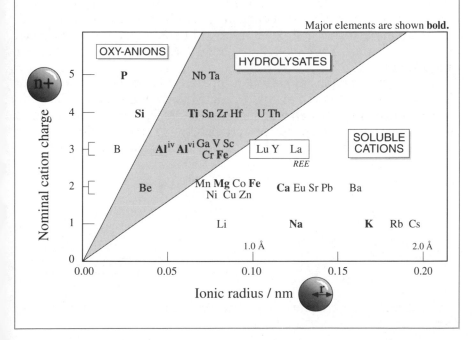

Major elements are shown **bold.**

207

9.6.1 Organic carbon

In a book of this size, one can only scratch the surface of a subject as complex as organic geochemistry. A good summary written from the geologist's point of view is given by Krauskopf (1995); a fuller introduction can be found in the opening chapters of the book by Killops and Killops (1993).

The names of organic compounds used here conform to modern systematic nomenclature. Names given in brackets are the corresponding traditional names, which may be more familiar to some readers.

HYDROCARBONS

Hydrocarbons are compounds of carbon and hydrogen alone. The simplest are the **alkanes** (paraffins), a family of polymers related to methane:

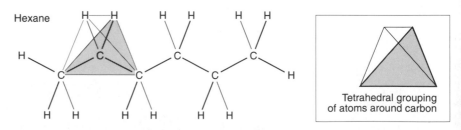

methane   ethane   propane

Many alkanes are chain-like polymers in which each carbon atom (except those at the ends of the chain) is attached to two neighbouring carbon atoms and two hydrogen atoms, but some alkanes have branch or ring structures. They have the general formula C_nH_{2n+2} (C_nH_{2n} for cycloalkanes).

Alkanes are examples of **saturated** organic compounds, consisting of molecules constructed entirely with single bonds. Though not shown in the simplified diagrams above, such molecules have a zigzag shape in three dimensions, owing the the tetrahedral (sp³-based) disposition of bonds around each carbon atom (Box 9.5). An example is hexane:

Hexane

Tetrahedral grouping
of atoms around carbon

Unsaturated compounds are those containing one or more double $C=C$ bonds or triple $C\equiv C$ bonds, like ethene (ethylene):

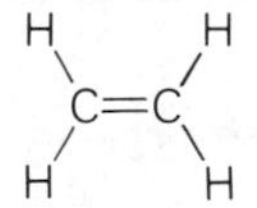

When forming double bonds (as we discovered with graphite – Box 7.4), carbon uses the sp²-hybrid, which has 120° 'planar trigonal' geometry. All of the nuclei in the ethene molecule therefore lie in the same plane.

The same is true of a large and important class of unsaturated organic compounds based on the **benzene** molecule C_6H_6:

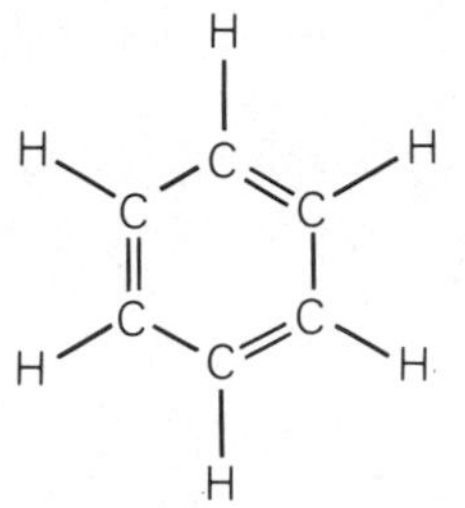

condensed notation (Box 9.5)

Compounds based on benzene, like styrene (from which polystyrene is manufactured) are called **aromatic** compounds, as a number of them have distinctive odours.

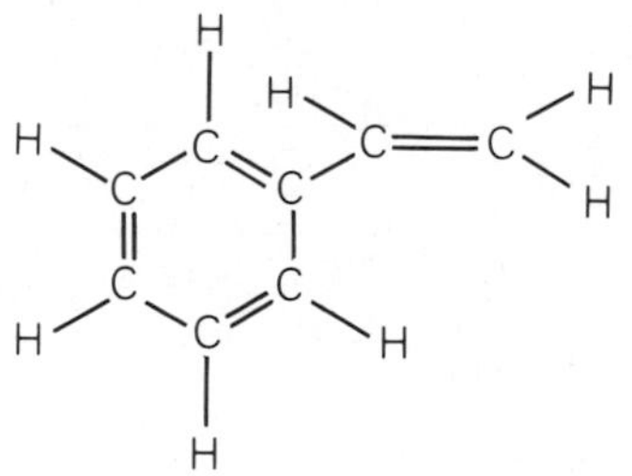

styrene

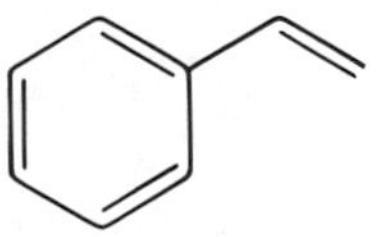

condensed notation (Box 9.5)

Open-chain (acyclic) and aromatic hydrocarbons are the main, and most useful, constituents of **petroleum** (Selinger, 1979, Chapter 11). Natural gas consists mainly of methane.

CARBOHYDRATES

Carbohydrates are a class of organic compounds in which C, H, and O combine in the proportion $C_x(H_2O)_y$, where x and y are integers. **Sugars** such as **glucose** ($C_6H_{12}O_6$)

Box 9.5

Notation for organic molecules

The complexity of most organic molecules necessitates a shorthand notation. Hydrocarbon molecules are often abbreviated to matchstick structures showing just the bonds between carbon atoms: at each angle in the symbol the appropriate CH_x group is implied.

The cyclic molecule of benzene is written as a hexagon incorporating three double bonds. This can be written in two ways:

The molecule actually adopts both configurations at once as with graphite (Box 7.4), the π-electron density is delocalized around the entire ring. This is sometimes symbolized as:

are the simplest carbohydrates. (Like alkanes, these are zigzag molecules in 3D.) Sugars consist of chains of a few of these C_6 units: sucrose, for example, is $C_{12}H_{22}O_{11}$. **Starches** consist of longer chains. The longest-chain carbohydrates are **celluloses** (with $x \sim 10\,000$). Cellulose is familiar as the main constituent of paper. Wood consists of about 70% cellulose, the remainder being an aromatic constituent called **lignin**, which gives wood its toughness and structural utility.

ACIDS, AMINO ACIDS AND PROTEINS

Acidic behaviour in organic molecules is associated with the presence of a **carboxyl** group, as illustrated by **ethanoic** (acetic) acid, familiar as the sharp-tasting constituent of vinegar:

carboxyl group

Groups of atoms like the carboxyl group, which involve the organic molecule to which they belong in certain specific types of reaction, are called **functional groups**. Another example is the **amino group** (NH_2), whose presence in the same molecule as the carboxyl group is the characteristic of a particularly important class of organic acids, the **amino acids**. About 20 amino acids occur in the living world. The simplest, glycine, can be recognized as a derivative of ethanoic acid:

amino group → carboxyl group

In solution, amino acids behave as weak acids: they dissociate by releasing a proton from the carboxyl group:

$$NH_2CH_2COOH \rightarrow H^+ + NH_2CH_2COO^- \qquad (9.1)$$

At the same time, however, the amino group has the capacity to accept a proton, thereby behaving as a base:

$$H^+ + NH_2CH_2COO^- \rightarrow {}^+NH_3CH_2COO^- \qquad (9.2)$$

Amino acids thus have the remarkable property of forming dipolar ions or **zwitterions** (a German term meaning 'hybrid ion') with opposite charges at each end. No doubt this contributes to their high solubility in polar solvents like water or ethanol, each end being solvated by a sheath of solvent molecules (Box 4.1).

The capacity of the amino acids to react with acids or bases according to circumstances means that molecules can also react with each other. Reaction between the amino group of one molecule and the carboxyl group of another joins the two molecules together:

211

Nature uses this **peptide linkage** reaction to assemble amino acid units into huge **protein** molecules, whose relative molecular masses run into thousands. Proteins are the essential constituents of the living cell.

Proteins in decaying organic matter decay rapidly by **hydrolysis** (Box 9.4), the reverse of the reaction above, leading to breakdown into simpler proteins and amino acids.

9.6.2 Inorganic carbon

The element carbon exists in the Earth in two crystalline forms, graphite and diamond (whose structures and properties are compared in Chapter 7). Diamond crystallizes only at very high pressure, equivalent to depths greater than 120 km (Box 2.2). The derivation of diamond from deep in the mantle, and geochemical evidence for CO_2-rich fluids in peridotite nodules brought up from similar depths, indicate that carbon is an important minor constituent of the mantle. Carbonaceous material is a major component of many meteorites (Box 10.3).

Carbon occurs in the crust chiefly as carbonate, most of which is biogenic. Limestones make up about 25% of the total mass of Phanerozoic sedimentary rocks.

CARBON DIOXIDE

In the **carbon dioxide** molecule each oxygen is doubly bonded to carbon. The carbon atom allocates two valence electrons to an **sp^1-hybrid**, whose lobes stick out in opposite directions, and the σ-bonds they form give the molecule a linear O–C–O shape. The two carbon 2p

Box 9.6

Chlorophyll

We have seen the key role that nitrogen plays in the synthesis of proteins. Another compound essential to life in which nitrogen is a vital constituent is **chlorophyll**, the green pigment in plants upon which photosynthesis primarily depends. Chlorophyll is actually a class of closely related compounds that share the following characteristics:

(a) The fundamental architecture comprises four molecules of pyrrole (C_4H_5N), linked together to form a ring structure known as a porphyrin (see Box 9.5 for symbolism):

Porphyrin

Pyrrole

(The porphyrin ring structure is found in other biologically essential compounds such as haemoglobin and vitamin B_{12}.)

(b) At the centre of the porphyrin ring lies a magnesium atom, forming a co-ordination complex with the surrounding pyrrole nitrogens.

(c) Various alkyl and carboxyl groups are attached to the porphyrin ring, whose identities differ with the variant of chlorophyll.

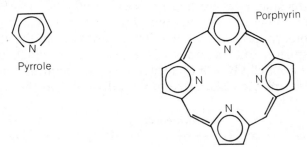

Chlorophyll a

C=O

COOCH$_3$

COOC$_{20}$H$_{40}$

The alternation of double and single bonds in the porphyrin ring resembles that in graphite (Box 7.4) and benzene, and in a similar way the bonds coalesce to form interconnecting molecular orbitals above and below the ring. Chlorophyll absorbs light strongly at the blue and red ends of the visible spectrum (hence its green colour – Figure 8.3), photochemically releasing electrons from these orbitals and setting in train a complex set of biochemical reactions by which CO_2 is ultimately reduced to carbohydrate (reaction 9.2).

orbitals not involved in the hybrid form π-bonds with the oxygen atoms on either side.

CO_2 is an acidic oxide (Figure 9.4). On dissolving in water it forms a slightly acidic solution (Chapter 4). Rain water in equilibrium with atmospheric CO_2 has a pH of 5.7, which partly accounts for its effectiveness as a weathering agent. In the northern hemisphere, however, this acidity is reinforced by oxides of sulphur and nitrogen introduced into the atmosphere by the burning of fossil fuels.

CO_2 is believed to have been the principal constituent of the Earth's primordial atmosphere, as we find for Venus today ($CO_2 = 96\%$ of the Venusian atmosphere). Throughout most of the Earth's history, however, photosynthesis has 'drawn down' carbon from the atmosphere into the biosphere. Photosynthesis is a reaction by which plants and algae use chlorophyll (Box 9.6) to generate the reduced carbon they need from atmospheric CO_2, or from CO_2 dissolved in the oceans:

$$x H_2O + x CO_2 \xrightarrow[\substack{\text{absorbed by} \\ \text{chlorophyll}}]{h\nu \text{ (solar energy)}} (-CH_2O-)_x + x O_2 \qquad (9.3)$$

$$\substack{\text{carbohydrate} \\ \text{(simplified)}}$$

Aquatic biota (zooplankton and higher organisms) have also fixed carbon in carbonate shells which have accumulated as limestone. These processes have together reduced the CO_2 content of the Earth's atmosphere today to a few hundred parts per million.

The pre-industrial CO_2 level was about 280 ppm. This represents a biologically mediated balance between photosynthesis and respiration (which converts O_2 to CO_2). CO_2 is a 'greenhouse' gas; were it not present at all, the mean atmosphere temperature at the Earth's surface would be $-18°C$, but CO_2 and the other natural greenhouse gases (e.g. water vapour) warm the atmosphere by about $33°C$ and make the Earth habitable. For the past two centuries, however, the concentration of CO_2 has been rising as a consequence of fossil-fuel burning and deforestation, and the current level is in the region of 350 ppm and rising (Figure 9.2). The hemispheric balance between combustion and photosynthesis shifts with the seasons, to the left in winter and to the right in summer. Current estimates suggest that the 'anthropogenic greenhouse effect' may be warming the Earth by up to $0.5°C$ per decade, and this may have serious climatic and social consequences if immediate action is not taken to stem the rise. The role of CO_2 and other greenhouse gases in global warming is dealt with in detail by Nisbet (1991).

Carbon also forms a **monoxide** (CO) which occurs in volcanic gases and in the atmosphere, but at much lower concentrations than CO_2.

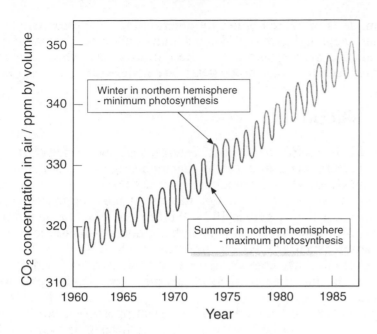

Figure 9.2 Annual and secular (non-cyclic) variation in atmospheric CO_2 content measured at the Mauna Loa observatory in Hawaii (chosen for its remoteness from centres of industrialization).

9.6.3 Carbon isotopes

There are two stable isotopes of carbon, ^{12}C ($\sim 98.9\%$) and ^{13}C ($\sim 1.1\%$). Owing to the 8% mass difference in relative atomic mass, geochemical reactions discriminate to a small extent between the two isotopes. Dissolved carbon dioxide and carbonate sediments in the oceans contain a higher $^{13}C/^{12}C$ isotope ratio (by 5–10 parts per thousand) than atmospheric CO_2. Photosynthesis, on the other hand, brings about the reverse fractionation: organic matter is quite strongly depleted in ^{13}C, leading to $^{13}C/^{12}C$ ratios in – for example – marine phytoplankton that are about 20 parts per thousand lower than for atmospheric carbon dioxide. This makes the $^{13}C/^{12}C$ ratio a valuable tracer for photosynthesis in the geological record, for instance in detecting the beginnings of life in ancient sedimentary rocks.

The short-lived radioactive carbon isotope ^{14}C is important as a dating tool in archaeology and Quaternary geology. Terrestrial ^{14}C (half-life 5730 years) is formed by the bombardment of ^{14}N nuclei in the upper atmosphere by cosmic rays. The radiocarbon atoms formed become part of the atmospheric inventory of CO_2, and soon enter plant tissue through photosynthesis. Plants maintain carbon-isotope equilibrium with atmospheric CO_2 during their lifetime, but this exchange ceases when the plant dies. The ^{14}C trapped in dead vegetable matter decays with time, providing an isotopic clock applicable to shorter intervals of time (thousands and tens of thousands of years) than the K–Ar and Rb–Sr dating methods.

In **radiocarbon dating**, as this technique is called, the ^{14}C remaining in a sample is measured either by determining its radioactivity, or by counting the number of ^{14}C atoms individually with the aid of an ultra-sensitive mass spectrometer. Recent advances, which extend the technique back to about 50 000 years, are reviewed by Hedges (1985).

9.7   Silicon

Silicon (Si) is a hard metalloid of intermediate electronegativity (1.9), with a structure identical to that of diamond (Figure 7.5). Like the next element in Group IV, germanium (Ge), it has become very important as a semiconductor (Chapter 7). For this use it must be extremely pure (impurities less than 1 part in 10^{10}). 'Silicon' the element should not be confused with **silicone**, a class of synthetic **organo-silicon** polymers, in which groups such as CH_3 are attached to $-Si-O-Si-O-Si-$ chains and networks. Such compounds are widely used as lubricants and insulators, having greater thermal stability than equivalent organic polymers.

Silicon is the most abundant of the electropositive elements in the Earth's crust. It invariably occurs in the oxidized state (valency 4), as SiO_2 or silicate polymers (see p. 176).

SiO_2 occurs in a variety of structural forms, both crystalline and amorphous. As well as forming megascopic crystals, quartz occurs commonly in the **crypto-crystalline** form chalcedony, familiar as varieties like agate, jasper, chert and flint. The only truly amorphous form of silica is opal.

Quartz has a low but significant solubility in water of about 6 ppm at room temperature. SiO_2 solubility increases markedly with temperature, providing a geothermometer that can be used to estimate deep temperatures in hot springs.

Dissolved silica exists in the hydrated form $Si(OH)_4 = H_4SiO_4$, known as silicic acid (analogous to carbonic acid but weaker). Except near to ocean-floor hot springs and lava eruptions, sea water is undersaturated with silica everywhere. Diatoms and radiolaria are nevertheless able, by extracting SiO_2 from sea water, to secrete shells of opaline silica. They do so mainly in the uppermost photic zone of the oceans, where sunlight promotes a high biological productivity. Such biogenic precipitation of silica dramatically reduces the level of dissolved silica in the surface layer, to the extent that the available silica actually controls the populations of these organisms; Si is thus an example of a **biolimiting** element. The siliceous hard parts these organisms secrete dissolve relatively slowly and therefore accumulate on the ocean floor, eventually to be lithified into flinty rock called **chert**, which may often be seen under the electron microscope to consist of radiolarian debris. Some cherts, however, are abiogenic in origin.

9.8   Nitrogen and phosphorus

Nitrogen is familiar as the unreactive diatomic gas N_2 making up the major part of the atmosphere (Figure 9.3). The electronegative nitrogen atom has three vacancies in the valence shell, allowing three covalent bonds (one σ-bond and two π-bonds) to be established between the two atoms in the molecule.

Nitrogen adopts a range of valency states. It forms three stable gaseous oxides: nitrous oxide (N_2O), nitric oxide (NO) and nitrogen dioxide (NO_2). 'NO_x' is a convenient abbreviation covering all three. Significant amounts of NO and NO_2 are produced during combustion of fossil fuels, notably by cars; they contribute to the formation of the photochemical smog that threatens air quality in many large cities on hot summer days (Fergusson, 1982, Chapter 9; O'Neill, 1992, pp. 104–7).

Nitrogen is an important constituent of all living matter: the importance of the amino group ($-NH_2$) and amino acids has been discussed under carbon. In decaying organic matter, such compounds are decomposed bacterially to the gas ammonia (NH_3), most of which is oxidized by soil bacteria to nitrate (NO_3^-), the form of nitrogen most

B	C	N
Al	Si	P
Ga	Ge	As
In	Sn	Sb

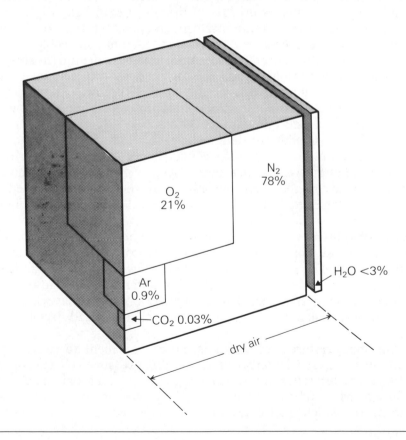

Figure 9.3 The composition (volume percentages) of the atmosphere at sea level.

readily utilized by plants. A large amount of nitrate is used as fertilizer which, owing to its high solubility, is washed into streams, rivers and lakes where it may give rise to serious pollution.

Phosphorus, unlike nitrogen, does not occur in the elemental state in nature. In silicate analyses (Box 8.3), it is reported as the oxide P_2O_5. It exists geologically as the phosphate oxy-anion (PO_4^{3-}), most commonly in the accessory mineral **apatite** ($Ca_5(PO_4)_3OH$). In basic magmas, phosphorus behaves as a high field-strength **incompatible element** (Box 9.1), but as crystallization proceeds such melts eventually become saturated with apatite, whose crystallization then leads to the depletion of phosphorus in subsequent melt fractions.

9.9 Oxygen

B	C	N	O
Al	Si	P	S
Ga	Ge	As	Se
In	Sn	Sb	Te

Viewed from a terrestrial perspective, oxygen is the most important of all chemical elements, being the most abundant element in the crust and mantle, and a major life-supporting constituent of the oceans and atmosphere (Figure 9.3).

Oxygen ($1s^2 2s^2 2p^4$) is the second most electronegative element (3.4, Figure 6.3). It is divalent, having two vacancies in the valence shell. Diatomic oxygen O_2 makes up 21% of dry air at sea level. At altitudes between 10 and 60 km, in the stratosphere, another type of oxygen molecule plays a significant rôle: this is the tri-atomic molecule **ozone**, O_3. Though the O_3 concentration rarely exceeds 10 ppm it absorbs ultraviolet radiation strongly, and the 'ozone layer' protects terrestrial life from the damaging effects of the Sun's UV rays. At sea level, however, ozone is an undesirable pollutant, contributing to photo-chemical smog and acid-rain damage (O'Neill, 1992).

Oxygen forms an oxide (oxidation state $-II$) with almost every other element. Oxides may have basic or acidic properties (Box 8.1), or they may be amphoteric, exhibiting both aspects. Because these characteristics depend on the electronegativity of the element bonded to oxygen, they correlate with its position in the Periodic Table (Figure 9.4).

The availability of oxygen determines the stability of many minerals, particularly those containing elements like iron that have multiple oxidation states. The availability of oxygen in low-temperature, aqueous environments is expressed by the **oxidation potential** Eh (Figure 4.1b); environments with free access to atmospheric oxygen have high Eh values, whereas anaerobic conditions are characterized by low Eh values.

In high-temperature systems, it is more convenient to express the availability of oxygen in terms of its **partial pressure** (p. 81), or the related parameter called the **oxygen fugacity**, f_{O_2}. Partial pressure is appropriate as a measure of concentration only in low-pressure gas mixtures, in which molecules are so dispersed that they behave

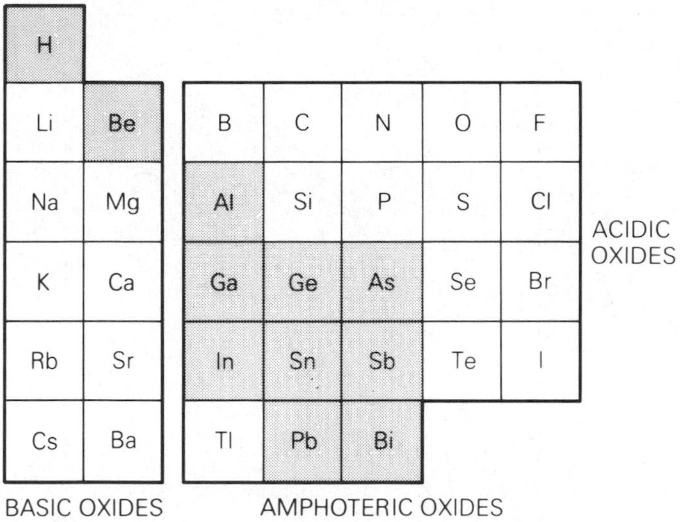

Figure 9.4 Acidic and basic oxides of the non-transition elements.

independently of each other (except when colliding). This state is called a 'perfect gas'. In high-pressure mixtures or when a gas is dissolved in another phase such as an igneous melt, gas molecules interact more strongly with neighbouring molecules, rather like ions in non-ideal aqueous solutions. Oxygen fugacity f_{O_2} is analogous to **activity** in describing the 'effective concentration' of oxygen in these non-ideal conditions.

Consider the reaction between iron-rich olivine crystallizing from a magma and oxygen dissolved in the melt:

$$3Fe_2SiO_4 + O_2 \rightleftharpoons 2(FeO.Fe_2O_3) + 3SiO_2 \qquad (9.4)$$

$$\underset{\substack{\text{\textit{fayalite}}\\\text{(\textit{olivine})}}}{}\quad \underset{\text{\textit{melt}}}{}\qquad \underset{\text{\textit{magnetite}}}{}\qquad \underset{\text{\textit{quartz}}}{}$$

Fayalite is a ferrous (Fe^{2+}) compound, whose crystal structure tolerates only a minute amount of Fe^{3+}. If sufficient oxygen is present in the system to cause significant oxidation of Fe^{2+} to Fe^{3+}, the olivine breaks down to a mixture of magnetite ($Fe_3O_4 = FeO.Fe_2O_3$), which contains both Fe^{2+} and Fe^{3+}, and quartz.

The reaction between fayalite and oxygen occurs at a specific $f_{O_2}–T$ conditions, as shown in Figure 9.5 (a sort of phase diagram). A univariant **equilibrium boundary** separates fields where fayalite + oxygen and magnetite + quartz are the stable assemblages. The reaction can proceed in either direction, and the coexistence of all four phases – as recorded by coexisting phenocrysts of fayalite, magnetite and quartz in a fine-grained volcanic rock, for instance – suggests that crystallization occurred under conditions lying somewhere on the equilibrium boundary. Because this is a univariant equilibrium (cf. Figure 2.2), one

219

Figure 9.5 f_{O_2}–T equilibrium diagram showing the experimentally determined reaction boundaries for the magnetite (Mt)–hematite (Hm) and fayalite (Fa)–magnetite–quartz(Q) reactions.

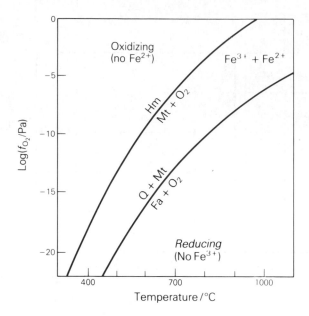

would need to estimate the temperature of crystallization by some other means before a numerical value of f_{O_2} could be worked out.

Under more oxidizing conditions, as shown by the upper equilibrium boundary in Figure 9.5, the Fe^{2+} in magnetite becomes oxidized, causing it to recrystallize as the mineral hematite (Fe_2O_3), in which the iron is entirely ferric.

9.9.1 Oxygen isotopes

Oxygen consists of three isotopes, ^{16}O, ^{17}O, and ^{18}O. In common with the stable isotopes of hydrogen (^{1}H, ^{2}H) and carbon (^{12}C, ^{13}C), precise measurement of the relative proportion of the oxygen isotopes shows minute variations between different oxygen-bearing phases that can be used as an **isotopic tracer** for analysing various geological processes (Box 9.7).

9.10 Sulphur

Native sulphur (oxidation state 0) forms yellow encrustations around volcanic vents and fumaroles, where it crystallizes as a **sublimate**. It can also be deposited from hot springs rich in H_2S or SO_2 and can occur in sedimentary rocks as a result of bacterial reduction of sulphate.

Sulphur can form compounds either with elements less electro-negative than itself (hydrogen and the metals) or with oxygen, which is more electronegative. It is useful to distinguish these two tendencies as 'reduced sulphur' and 'oxidized sulphur' respectively.

9.10.1   Reduced sulphur compounds

Sulphur is an essential nutrient element for living things, and organo-sulphur compounds therefore have considerable biochemical importance. They produce the distinctively pungent flavour and odour of onions and garlic. Hydrogen sulphide (H_2S), familiar as the smell of rotten eggs, is produced by decay of organic matter in anaerobic conditions, for example in stagnant water. Significant amounts of organo-sulphur compounds are present in oil, natural gas and coal; oxidation of these compounds to SO_2 during fuel combustion is the main anthropogenic cause of acid rain.

H_2S is also a significant constituent of volcanic gases. This suggests that **sulphide** (oxidation state $-II$) is the predominant form of sulphur in the Earth's interior. A great many metals of economic importance are deposited from hydrothermal fluids in the form of sulphide minerals (Box 9.8). The other Group VI elements, **selenium** (Se) and **tellurium** (Te), may take the place of sulphur in such minerals. A number of selenide and telluride minerals are known.

9.10.2   Oxidized sulphur compounds

Sulphides are not stable in contact with atmospheric oxygen. H_2S is rapidly oxidized, in water to the sulphate anion SO_4^{2-} (if sufficient dissolved oxygen is present) and in air to gaseous sulphur dioxide (SO_2, oxidation state IV). About 10^8 tonnes per year of SO_2 are released into the atmosphere by fossil-fuel burning and other industrial processes. On dissolving in water droplets (Box 9.4), SO_2 oxidizes to form an aerosol of sulphuric acid (H_2SO_4, oxidation state VI). This is the main contaminant in **acid rain**, which is causing increasingly severe and widespread damage to lakes, rivers and forests in the northern hemisphere.

Sulphide minerals are also susceptible to atmospheric oxidation. The weathering of near-surface sulphide ore bodies by the downward percolation of oxygenated water may lead to further enrichment of the ore. Just below the surface (Figure 9.6), sulphides are oxidized to sulphates, which migrate downward in solution. Commonly a number of carbonate, sulphate and oxide ores are precipitated just above the water table, whereas reducing conditions below the water table lead to deposition of **secondary** sulphide minerals. This is called **supergene** enrichment.

Sulphate minerals occur in two other environments. Barite ($BaSO_4$) occurs in low-temperature hydrothermal veins (as in the English Pennine orefield – Chapter 4), commonly in association with sulphides. Minerals like anhydrite ($CaSO_4$) and gypsum ($CaSO_4.2H_2O$) are characteristic of evaporites.

Box 9.7

Stable isotope tracers: oxygen and hydrogen isotopes

Oxygen, like hydrogen, carbon and sulphur, consists of a number of isotopes whose relative atomic masses differ by several percent:

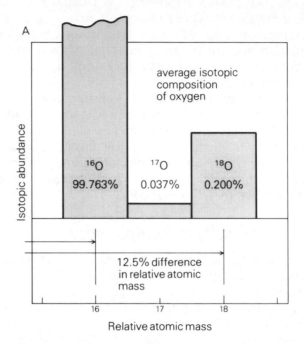

A

average isotopic
composition
of oxygen

Isotopic abundance

^{16}O
99.763%

^{17}O
0.037%

^{18}O
0.200%

12.5% difference
in relative atomic
mass

16   17   18

Relative atomic mass

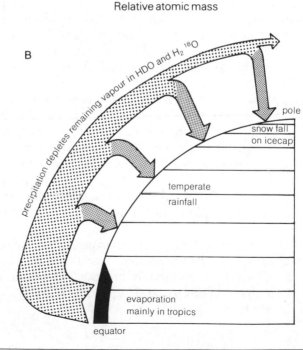

B

precipitation depletes remaining vapour in HDO and $H_2\,^{18}O$

pole
snow fall
on icecap

temperate
rainfall

evaporation
mainly in tropics

equator

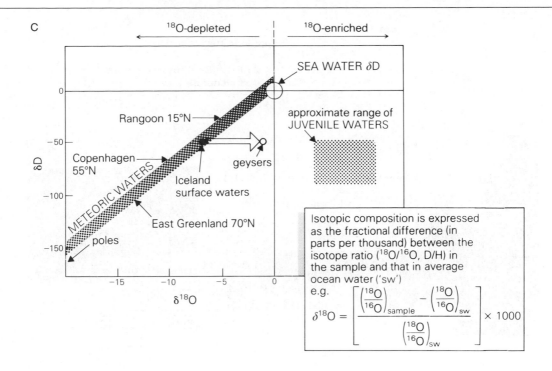

The $H_2{}^{18}O$ molecule is 11% heavier than $H_2{}^{16}O$, which makes it very slightly more difficult to evaporate. Its vapour pressure at 100 °C is 0.5% lower than that of $H_2{}^{16}O$, causing its boiling point to be 0.14 °C higher. Vapour in equilibrium with water will be slightly deficient in $H_2{}^{18}O$ and HDO ($^1H^2H^{16}O$) relative to the coexisting liquid. Because atmospheric water vapour is produced by evaporation of sea water, it is slightly depleted in these heavier molecules. Furthermore, precipitation of rain makes the remaining vapour even more depleted (diagram B).

Precipitation becomes progressively more depleted in $H_2{}^{18}O$ and $HD^{16}O$ (a) with increasing latitude (diagram C), and (b), in a continental area, with increasing distance from the ocean.

H and O isotope ratios enable us to distinguish three broad categories of water which can be involved in geological reactions (see diagram C):

(a) Sea water. (When trapped in rocks this is called **connate** water.)
(b) Rain-derived (**meteoric**) ground water, having variable negative δD and $\delta^{18}O$ (see above).
(c) Magmatic and metamorphic water in isotopic equilibrium with silicate rocks at high temperatures (**juvenile** water), with positive $\delta^{18}O$ but negative δD.

The isotopic composition of waters may however be modified by interaction with hot rocks, making their ultimate origin less clear (e.g. geysers).

The water participating in a geological process usually leaves an isotopic fingerprint on that products of that process. Oxygen isotope analysis of a continental hydrothermal ore deposit, for example, will often reveal negative δ values that indicate a meteoric origin for the hydrothermal fluid.

Box 9.8

Sulphide minerals and 'soft' metals

Not all metals exhibit the capacity to form sulphide minerals. The elements that do, known among geochemists as **chalcophile** elements (Figure 10.3), mostly lie in the right-hand side of the d-block or the adjacent portion of the p-block.

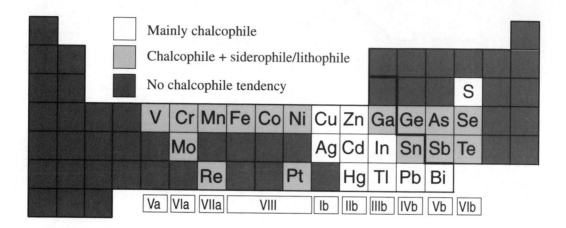

Why is chalcophile behaviour restricted to such a specific region of the Periodic Table? Chemists divide metal ions into 'hard' and 'soft' Lewis acids (Box 7.5). 'Hardness' in this sense is a characteristic of the strongly electropositive metals on the left-hand side of the Periodic Table (e.g. the alkali and alkaline earth metals) which form very ionic bonds, particularly with strongly electronegative elements like oxygen. Chalcophile metal ions on the other hand behave as 'soft' acids: they are well endowed with d-electrons and have high electronegativities (1.7–2.5, Figure 6.3); their ions are easily polarized and therefore as metals go they form relatively covalent bonds. Such bonds can be established most effectively with ligands of low electronegativity such as sulphide (which is classed as a 'soft base'). The tendency for hard acids (Na^+, K^+, Mg^{2+}, Ca^{2+}) to combine with hard bases (O^{2-}), whereas soft acids (e.g. Cu^{2+}, Ag^+, Hg^{2+}) associate with soft bases (S^{2-}), is a fundamental distinction in modern inorganic chemistry, corresponding closely to the geochemical division of elements into lithophile and chalcophile groups (Figure 10.3).

We have seen that bonding in sulphides has affinities with bonding in metals (Figure 7.8b). It is therefore not surprising to find that many sulphide minerals exhibit metal-like features, such as metalloid lustre, opacity and relatively high thermal and electrical conductivity. These arise from the low electronegativity contrast between the soft acid and soft base, the mean electronegativity of PbS (2.45), for example, being little different from the metal itself (2.3). In some Fe, Co and Ni sulphides the metallic character is further enhanced by direct metal-metal bonding, an intervening conduction band being formed by interaction between d-orbitals of neighbouring metal atoms.

Zn has the lowest electronegativity of the truly chalcophile elements, and sphalerite (ZnS) shows little metallic character; iron-free samples are translucent.

Sulphur commonly combines with metals in non-integer proportions. An example of such a **non-stoichiometric** sulphide is pyrrhotite, whose composition is best represented by the formula $Fe_{1-x}S$, where x can lie between 0.0 and 0.15.

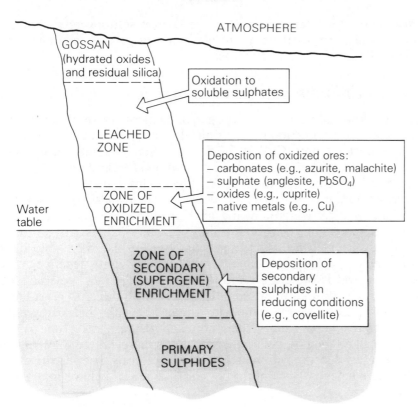

ATMOSPHERE

GOSSAN
(hydrated oxides
and residual silica)

Oxidation to
soluble sulphates

LEACHED
ZONE

Deposition of oxidized ores:
– carbonates (e.g., azurite, malachite)
– sulphate (anglesite, $PbSO_4$)
– oxides (e.g., cuprite)
– native metals (e.g., Cu)

Water
table

ZONE OF
OXIDIZED
ENRICHMENT

ZONE OF
SECONDARY
(SUPERGENE)
ENRICHMENT

Deposition of
secondary
sulphides in
reducing conditions
(e.g., covellite)

PRIMARY
SULPHIDES

Figure 9.6 Idealized cross-section of a near-surface sulphide ore body, showing the zonation due to percolating oxygen-bearing solutions. The mineral assemblages produced are illustrated by typical copper minerals (whose stability relations are indicated in Figure 4.1a), but many other minerals occur in such environments.

9.11   Fluorine

Fluorine has the highest electronegativity of all elements, and is the most reactive. It forms strongly ionic compounds. The commonest fluoride mineral (and the chief industrial source of fluorine) is **fluorite**, CaF_2 (Figure 7.3d), which occurs most commonly in hydrothermal veins. The ionic radius of the fluoride anion F^- (1.25 Å, Box 7.2) is similar to those of O^{2-} and OH^-, and fluorine is a common substituent for OH^- in hydrous minerals like amphiboles, micas and apatite.

Being more reactive and electronegative, fluorine can displace oxygen from most silicates. Dissolving hydrogen fluoride (HF, a gas) in water produces **hydrofluoric acid**, which is widely used in analytical geochemistry as it is the only acid capable of attacking silicate rocks (in powdered form) to bring them into solution for analysis. It is a dangerous reagent that requires special training and handling: unlike the more familiar hydrochloric acid it causes no burning sensation on contact with the skin, but penetrates into deeper tissues and causes

					He
B	C	N	O	**F**	Ne
Al	Si	P	S	**Cl**	Ar
Ga	Ge	As	Se	**Br**	Kr
In	Sn	Sb	Te	**I**	Xe

225

intense pain after a few hours. Because HF (as solution or gas) attacks glass, it must be used only in platinum or plastic containers, in specially designed fume cupboards.

9.11.1 Chlorine and other halogens

Chlorine is the third most electronegative element. In silicate rocks it is a trace element, being about four times less abundant in the crust than F. The chloride anion (Cl^-) is, however, the most abundant dissolved species in sea water, and it is the dominant **ligand** in brines and hydrothermal fluids. Chlorine gas (Cl_2) has many industrial uses.

Organochlorine compounds are widely used in industry as solvents, propellants and refrigerants owing to their extreme chemical inertness. Among the most inert are chlorofluorocarbons (CFCs), hydrocarbon derivatives in which every hydrogen has been replaced by chlorine or fluorine atoms. Whereas elemental chlorine is highly reactive in the atmosphere and is therefore rapidly washed out by rain, volatile CFCs may persist for up to a hundred years. They are powerful 'greenhouse' gases, but pose a more immediate threat through their capacity to deplete stratospheric ozone: their atmospheric residence time is long enough to allow them to penetrate the stratosphere, where they undergo photochemical decomposition and release Cl radicals that destroy ozone.

The rarer halogens, bromine (Br) and iodine (I), behave in a similar way to chlorine.

9.12 Noble gases

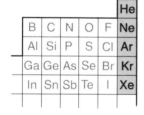

The elements helium (He), neon (Ne), argon (Ar), krypton (Kr) and xenon (Xe) have completely filled valence shells. This electronic structure is so stable that these elements display negligible chemical reactivity, and exist (except at extremely low temperatures, Box 7.7) as monatomic gases. They are known as the **noble**, **inert** or **rare gases**.

Though helium is the second most abundant element in the universe after hydrogen (Figure 10.2), it constitutes only 0.000 52% of the Earth's atmosphere. Unlike other gaseous elements like N_2, helium has not accumulated in the atmosphere because its relative atomic mass is too low for it to be retained by the Earth's gravitational field: the escape velocity of helium atoms (and H_2 molecules) is well below the actual average thermal velocity of such atoms at normal temperatures.

Helium consists of two stable isotopes, ^{3}He and ^{4}He. In natural helium escaping through the Earth's surface (recovered from oilfield brines or hot springs, for example), ^{3}He is a relic of the He originally incorporated during the accretion of the Earth (Chapter 10), whereas ^{4}He – about a million times more abundant – is almost entirely the product of the radioactive decay of thorium (Th) and uranium (U) throughout the Earth's history.

The principal interest in Ar is for radiometric dating, using the K−Ar method (Box 9.2) or the more recently developed $^{39}Ar-^{40}Ar$ method (Faure, 1986).

9.13 Transition metals

Most metals that we use in the home, the office and in industry are **transition metals**, making up the d-block of the Periodic Table. The essential feature of a transition metal is the presence in the atom or ion of a **partially filled d sub-shell** (Chapter 6). The d-block can be divided into first, second and third **transition series** (Figure 9.7), according to

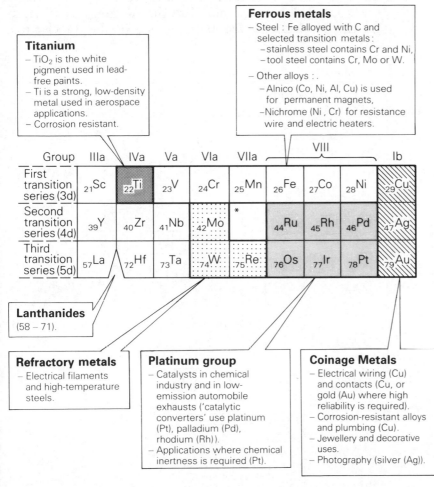

Figure 9.7 Transition metals and their uses.

*Technetium (Tc, $Z = 43$) has no stable isotope. Not found in nature.

227

whether the 3d, 4d or 5d sub-shell is the partially filled sub-shell. The elements of Group IIb (zinc, cadmium and mercury), as they have full d sub-shells, are not regarded as transition metals.

The transition metals share a number of important chemical characteristics:

(a) Most transition metals and their alloys are tough, **chemically stable** metals that have innumerable industrial and domestic uses (Figure 9.7).

(b) Most can utilize more than one **oxidation state** (valency) in geological environments. The most familiar example is iron (Box 4.6). Figure 9.8 shows the range of oxidation states among the first transition series. In the elements up to manganese, all d-electrons are able to participate in bonding and high oxidation states can therefore be attained. In elements like iron, cobalt (Co) and nickel (Ni), the d sub-shell behaves more like the electron core (Chapter 5): only the 4s electrons and perhaps one of the 3d electrons

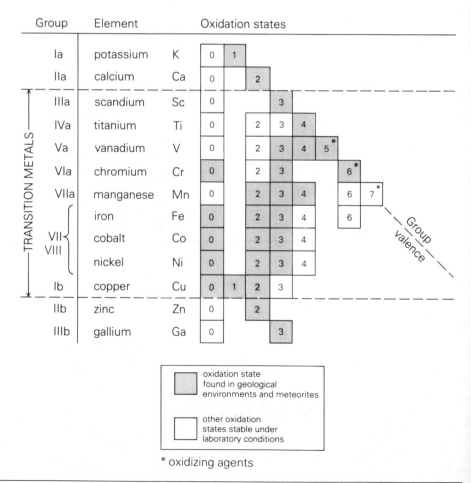

Figure 9.8 Oxidation states of the first transition series.

have energies high enough to be used in bonding, and only low oxidation states occur.

(c) Transition metals form a wide range of **co-ordination complexes**, some of which play an important part in stabilizing the metals in solution and promoting their transport in hydrothermal fluids (pp. 98, 158).

(d) Transition-metal compounds are often strongly **coloured** (Box 9.9). Many minerals owe their distinctive colours to the presence of a transition metal.

(e) Transition metals are responsible for the **magnetism** of minerals and rocks. This property, most prominent in the later members of the first transition series (Fe, Co, Ni), is due to the presence of **unpaired electrons** in the d sub-shell. Unlike valence electrons, these 3d electrons may remain unpaired when the metal atom combines in a compound. Paired electrons in an orbital generate equal and opposite magnetic fields which cancel out, but an unpaired electron causes a net magnetic field, which in the case of a few minerals like magnetite and pyrrhotite gives rise to permanent (remanent) magnetism.

9.14   Rare earth elements

Following the element lanthanum, La (the first member of the third transition series – see Figure 9.7), electrons begin to occupy the seven 4f orbitals, forming the fourteen metals known as the **lanthanides** or **rare earth elements** (**REE**; p. 202). The distinction between individual rare earths lies in the number of 4f electrons. These are mostly not involved in bonding, and the chemical properties of all 14 elements, together with lanthanum, are therefore remarkably similar. All have stable trivalent states (Figure 9.9).

Due to the increase in nuclear charge, there is a steady decrease of ionic radius from lanthanum La^{3+} to lutetium Lu^{3+} (the **lanthanide contraction** – Figure 9.9). The 'light rare earths' ('LREE', La–Sm) are incompatible elements. Owing to their smaller ionic radii, however, the 'heavy rare earths' ('HREE', Gd–Lu) are more easily accommodated in the crystal structures of some rock-forming minerals, particularly garnet and amphibole. The rare earth elements therefore provide the geochemist with an array of trace elements which, though virtually identical in other chemical properties, range continuously in behaviour from incompatible to selectively compatible.

For reasons discussed in Chapter 10, an element of even atomic number tends to be about ten times more abundant than its neighbours with odd Z-values, giving the rare earths in particular (Figure 10.2 inset) a characteristic zig-zag abundance pattern that is common to all Solar System matter. When examining the rare-earth pattern of a terrestrial or lunar rock, therefore, geochemists eliminate this sawtooth

Box 9.9

Transition metals and the colour of minerals

d-orbitals project a long way from the nucleus and are highly directional (Figure 5.5). d energy levels are therefore sensitive to the positions of surrounding anions. The diagram depicts a transition metal in a regular octahedral site in a crystal, surrounded by six equidistant anions, which one can imagine positioned on the reference axes used for describing orbital geometry (Chapter 5). The co-ordination structure is shown cut in half, to clarify the geometry. Because of the potential repulsion between these anions and electron density in the d-orbitals, the most stable d-electrons will be those in orbitals that interfere least with the octahedrally positioned anions.

Placing a transition metal in octahedral co-ordination brings two changes in d energy levels. The mean energy increases, due to the overall repulsion by the anion field. Secondly the energy levels are split. Orbitals like the d_{yz} example shown, whose lobes point at the edges of the co-ordination polyhedron between the anions), have a lower energy than orbitals pointing directly at the anions, which experience maximum repulsion.

The split in d energy levels (Δ_{oct}) varies with the identity of the cation and the crystal site. For many transition metals in minerals, the energy difference between the d levels corresponds to the photon energy of visible light. Such ions are therefore capable of absorbing strongly certain wavelengths in the visible spectrum, by promoting electrons from the lower level to a vacancy in the upper level (cf. Figure 6.4). This **crystal field splitting** is the cause of the strong colours of minerals like malachite and azurite (Cu) or olivine (Fe).

The presence of d-electrons effectively causes a transition metal ion to deviate significantly from the spherical shape assumed in Chapter 7. This affects the ease with which such an ion can be accommodated in a crystal site. Consider the case of nickel, an important trace element in basalts: Ni^{2+} in a basalt melt crystallizing olivine exhibits an unexpectedly strong preference for the octahedral Mg site in olivine, to such an extent that crystallization of olivine rapidly depletes the Ni content of the melt. The reason is because the olivine site more readily accommodates the d-orbital geometry of Ni^{2+} than do the available sites in the melt.

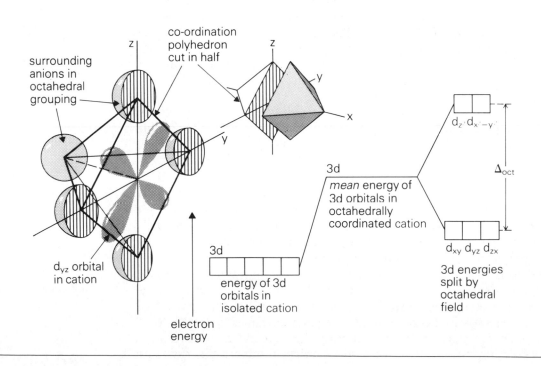

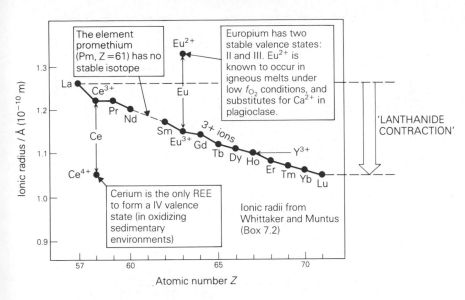

Figure 9.9 Ionic radii of the rare earth elements. Y^{3+} represents the related element yttrium (Figure 9.7).

effect by dividing each REE concentration in the rock (ppm) by the average concentration of the same element in chondrite meteorites (also ppm). Chondrites serve as a sensible reference value here because they provide a reasonable estimate of the primordial composition of the Earth's mantle, the 'starting point' from which all igneous rocks have ultimately been derived. The result is a smooth 'chondrite-normalized' REE pattern showing how much each rare earth has been 'enriched' in the sample of interest relative to a model mantle source.

Figure 9.10 shows a chondrite-normalized REE pattern for two lunar basalts. The lower pattern represents lava formed directly by melting of the lunar mantle; the pattern is relatively flat, suggesting that the lunar source region differed little from chondrite, at least in terms of REE abundances. The upper pattern is also flat in overall terms but differs in two important ways: REE concentrations are generally about ten times higher, but there is a pronounced deficiency in the abundance of the element europium (Eu), a 'negative Eu anomaly'. This basalt evidently represents magma that has undergone partial crystallization. Most REE are incompatible in basalt minerals, and crystallization of olivine, clinopyroxene, plagioclase, etc. must therefore concentrate REE in the declining amount of melt. The exception is europium which, unique among the REE, has a 2+ oxidation state (box in Figure 9.9). Eu^{2+} has a radius similar to Ca^{2+} (Box 9.1), and can substitute for Ca in plagioclase (though not in clinopyroxene, whose more compact structure excludes it). When plagioclase is crystallizing under relatively low f_{O_2} (reducing conditions), therefore, Eu may behave as a compatible element and become depleted in the melt as plagioclase crystallization removes it. Thus a negative Eu anomaly is a valuable geochemical tracer of plagioclase crystallization during magma fractionation.

231

Figure 9.10 Chondrite-normalized rare earth element diagram for two lunar basalts. Note that the vertical (enrichment) scale is logarithmic. REE patterns are prepared by plotting the ratio for each element, then joining up the array of points with straight lines. (Source: Taylor (1982), cited in Chapter 10.)

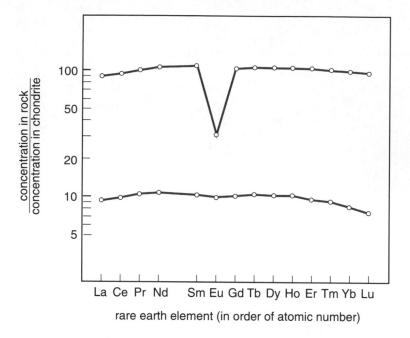

rare earth element (in order of atomic number)

REE are useful for detecting the influence of other minerals on igneous processes. For example, the presence of garnet in a basalt source region during melting causes HREE to be retained in the residual solid, leaving the melt formed relatively depleted in HREE relative to LREE. This rare earth pattern of such a basalt would show a steep negative slope.

The trace element yttrium (Y) also forms a 3+ ion. The ionic radius of Y^{3+} is the same as Ho^{3+}, and in geological materials yttrium is always closely associated with the HREE.

9.15 Bibliography

Faure, G. (1986) *Principles of isotope geology*, 2nd edn, Wiley, New York.

Fergusson, J. E. (1982) *Inorganic chemistry and the Earth*, Pergamon, Oxford.

Hedges, R. E. M. (1985) Progress in radiocarbon dating. *Science Progress*, **69**, 409–27.

Killops, S. D. and Killops, V. J. (1993) *An introduction to organic geochemistry*, Longman, Harlow.

Krauskopf, K. B. (1995) *Introduction to geochemistry*, 3rd edn, McGraw-Hill, New York.

Nisbet, E. G. (1991) *Leaving Eden*, Cambridge University Press, Cambridge.

O'Neill, P. (1992) *Environmental chemistry*, 2nd edn, Allen & Unwin, London.

Puddephatt, R. J. and Monaghan, P. K. (1986) *The Periodic Table of the elements*, 2nd edn, Oxford University Press, Oxford.

Selinger, B. (1979) *Chemistry in the market place*, John Murray, London.

9.16 Exercise

1 The reaction:

$$Fe_2(SO_4)_3 + 6H_2O \rightleftharpoons 2Fe(OH)_3 + 3H_2SO_4$$

is a balanced chemical equation. Adding up the total amount of any constituent on the left-hand side should equal that represented on the right-hand side: 2 atoms of Fe, 3 atoms of S, and so on. Below is a series of similar reactions (relating to supergene enrichment) which have not been balanced. Work out (by trial and error) the integers required in front of each molecule to produce a balanced equation.

(a) $Fe_2(SO_4)_3 + FeS_2 \rightarrow FeSO_4 + S$
(b) $CuFeS_2 + Fe_2(SO_4)_3 \rightarrow CuSO_4 + FeSO_4 + S$
(c) $FeSO_4 + O_2 + H_2O \rightarrow Fe_2(SO_4)_3 + Fe(OH)_3$
(d) $MnO_2 + H^+ + 2Fe^{2+} \rightarrow Mn^{2+} + H_2O + Fe^{3+}$
(e) $ZnS + Fe_2(SO_4)_3 + H_2O \rightarrow ZnSO_4 + FeSO_4 + H_2SO_4$

10
The elements in the universe

10.1 The significance of element abundance

Having considered the behaviour of some important elements in the lithosphere, hydrosphere and atmosphere (Chapters 4, 8 and 9), it is natural to ask the question: how were the chemical elements formed in the first place? Have they existed in their present form since the beginning of time, or has there been a progressive building up of the cosmic inventory? Can we identify the process(es) by which they were formed?

Current opinion favours the gradual synthesis of heavier elements from lighter ones, by a complex series of nuclear reactions occurring in stars. This process, called **stellar nucleosynthesis**, leaves its fingerprint on all forms of cosmic matter. We can learn how it works by studying the relative abundances of the elements in the universe as a whole, or in some representative part of it.

A second question we shall address in this chapter is how the Earth has come to have its present composition and structure. In seeking an answer to this problem, we again consider element abundances, noting this time how they differ between the Sun, the Earth and the other planets. Such differences provide clues to the kind of chemical processing that has produced the solar system and the Earth in their present form.

10.1.1 Elements and nuclides

An **element** consists of atoms that all share the same atomic number Z (the number of protons in the nucleus). The number of neutrons N may vary slightly: each individual value of N defines a different **isotope** of the element. Each isotope is identified by the value of the mass number $A = Z + N$, prefixing the element symbol: e.g. ^{90}Sr, ^{18}O, ^{14}C (Box 6.1). A **nuclide** is a hypothetical substance composed of atoms having particular values both of Z and N, i.e. a specific isotope of a specific element. The 81 stable elements comprise 264 stable nuclides (Box 10.1). Unstable – radioactive – ones are called **radionuclides**.

10.2 Measuring cosmic and solar system abundances

Our knowledge of the overall composition of matter in the Universe rests mainly on two kinds of analysis:

(a) spectral analysis of the light received by telescopes from stars, including the Sun, and from other radiant bodies such as nebulae (gas clouds);
(b) laboratory chemical analysis of meteorites, which represent the solid constituents of the solar system.

10.2.1 Spectral analysis

Stars are intensely hot nuclear fusion reactors. They derive their high temperatures and radiant output from the energy released when light nuclei (such as ^{1}H and ^{2}H) fuse together into more stable, heavier nuclei (such as ^{3}He). The theory of this **thermonuclear** process is outlined in Box 10.2.

In common with any very hot body, such as a red-hot poker or a light filament, the hot surface of a star radiates light consisting of a continuum of wavelengths (the 'white light' we receive from the Sun). Superimposed on this smooth electromagnetic spectrum are the absorption spectra of elements present in the cool outer atmosphere of the star, consisting of sharp dark lines, each of which can be associated with a specific chemical element. From these lines (**Fraunhofer lines**) the astronomer can establish the identity and abundance of most elements present. The 'calibration' factors used to translate absorption line density into element abundance have to be estimated theoretically, yet astronomers are confident that the abundance data available today, for about seventy elements in a great many stars, are mostly accurate to within a factor of two. As element abundances differ by as much as 10^{12} times (Figure 10.2), such uncertainties are tolerably small. Although stars progressively alter their element abundances through the nucleosynthesis taking place deep inside them, the cool outer envelope of a star is thought to remain representative of the material from which the star originally accreted.

In discussing the solar system, we shall be concerned with abundances in the Sun rather than stars in general.

Box 10.1

The nuclide chart

The clearest way to visualize the nuclides, both stable and radioactive, is to plot their Z values against N values as shown in the diagram.

The **stable** nuclides (solid circles) lie in a narrow, slightly curved ribbon. The most abundant light nuclides have $N = Z$, but heavier elements become increasingly more neutron-rich ($N \leqslant 1.5Z$). On either side of the stable-nuclide ribbon lies a band of radioactive nuclides, most of which are too short-lived to be found in nature. Neutron-rich radionuclides decay by ejecting a high-energy electron (for historical reasons called a beta-particle, and symbolized β^-):

$$\begin{array}{cccc} {}^{87}\text{Rb} & \rightarrow & {}^{87}\text{Sr} & + \beta^- + & \bar{v} \\ \textit{parent} & & \textit{daughter} & & \textit{antineutrino} \\ \textit{nuclide} & & \textit{nuclide} & & \end{array}$$

Loss of the beta-particle transforms one neutron in the nucleus into a proton, increasing Z by one at the expense of N (see inset diagram).

Proton-rich radionuclides decay either by ejecting a positron (β^+), or by capturing an orbital electron:

$${}^{40}\text{K} + e^- \rightarrow {}^{40}\text{Ar} + v \text{ (neutrino)}$$

by which a proton is transformed into a neutron ($Z \rightarrow Z - 1$, $N \rightarrow N + 1$). Both of these reactions are 'isobaric', leaving A unchanged (isobars lie on a diagonal line in Z, N-space — see diagram).

A third kind of nuclear decay is the ejection of an alpha-particle (2 protons + 2 neutrons = ${}^4\text{He}$ nucleus):

$${}^{147}\text{Sm} \rightarrow {}^{143}\text{Nd} + a^{2+}$$

Such a reaction reduces both Z and N by 2, and A by 4 (see inset diagram). The complicated decays of uranium and thorium isotopes to lead involve multiple a and β^- decays steps (Box 3.3).

${}^{87}\text{Rb}$, ${}^{40}\text{K}$, ${}^{147}\text{Sm}$ and uranium are all long-lived naturally occurring radionuclides which are used extensively in geo-chronology (Box 9.2).

Chart of atomic number Z versus neutron number N showing the naturally occurring nuclides

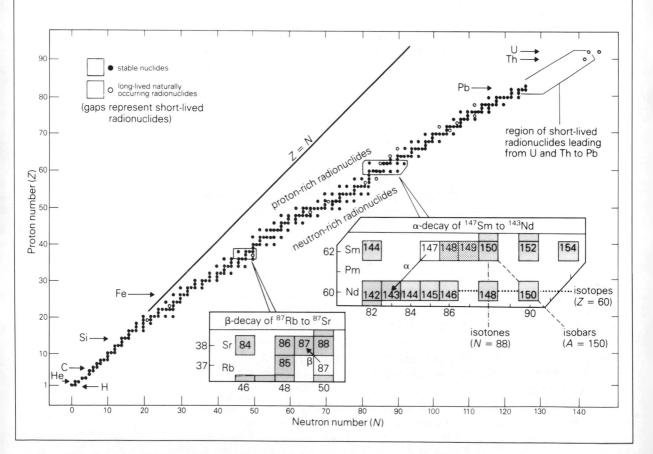

Box 10.2

Nuclear fusion and fission

Nuclei are held together by an immensely strong force called the **strong nuclear interaction** (SNI). It acts between nucleons only over very short distances similar to the size of the nucleus itself ($\sim 10^{-14}$ m). The more nucleons present in the nucleus, the stronger is the binding force that each one experiences. Counteracting the attractive force exerted by the strong nuclear interaction, however, is the electrostatic repulsion acting between the Z positively charged protons present, which, because the protons are held in such close proximity in the nucleus, is also an extremely powerful force.

The relative stability of nuclei can be expressed in terms of the mean potential energy per nucleon in the nucleus, relative to the potential energy a nucleon would possess as an isolated particle (set by convention at zero). Because every nucleus represents a more stable state than the same number of separate nucleons, the mean potential energy per nucleon is a negative quantity. Its variation with mass number A for the naturally occurring nuclides is sketched in the figure.

The shape of the graph reflects the interplay between the SNI and the electrostatic repulsion between protons. Where the curve drops steeply on the left-hand side the SNI is clearly the dominant force, but the curve flattens out around Fe (a region of maximum nuclear stability) and then rises gently as the proton-proton repulsion exerts a steadily more powerful influence; here the increase in SNI obtained by adding further nucleons to a nucleus is outweighed by the consequent increase in electrostatic repulsion.

Nuclei on the extreme left of the diagram, therefore, can in principle reduce their potential energy by fusing with other light nuclei to form heavier ones. **Fusion** of these lighter nuclei thus releases energy (it is exothermic) and this provides the source for the **thermonuclear** energy output of stars and hydrogen bombs. On the right of the diagram, on the other hand, is a region where fusion, were it to occur, would be energy-consuming (endothermic). Nuclei in this A range (>60) cannot be generated by fusion (see text). On the contrary, the heaviest nuclei, such as thorium and uranium, are radioactive and decay by emitting alpha-particles (Box 10.1; also Box 3.3): this is a mechanism for shedding mass and attaining a lower energy per nucleon (greater stability). The energy released by the decay of such elements within the Earth constitutes the largest component of terrestrial heat flow.

Certain heavy nuclides (^{235}U being the only naturally occurring example) are also **fissile**: on absorbing a neutron they split into two nearly equal nuclear fragments. These **fission products**, comprising various nuclides in the A range 100–150, have two important properties in common:

(a) They lie on a lower segment of the potential-energy curve than the parent nuclide. Thus fission is an exothermic process: it is the energy source for present nuclear-power reactors and for the original atom bomb.

(b) Although several neutrons are released in the fission process (which by colliding with other ^{235}U nuclei prompt further fission), the fission products still have higher $N:Z$ ratios than stable nuclides in the same A range, which makes them radioactive (Box 10.1). Their β-decay is responsible for the intense initial radioactivity of reactor wastes. (The longer-term radioactivty is due to α-emitting nuclides like the plutonium isotope ^{239}Pu.)

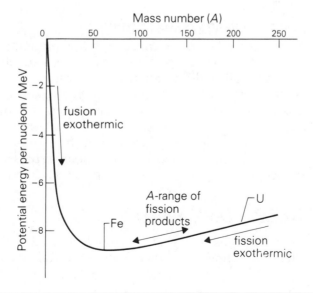

10.2.2 Analysis of meteorites

As anyone knows who has slept in the open on a clear starry night, shooting stars are a common phenomenon. They are the visible manifestation of hundreds of tonnes of solar-system debris that fall to Earth each day. Smaller infalling bodies may be vaporized completely in the atmosphere by frictional heating, but about 1% of them are large enough to survive as recoverable **meteorites**.

A simplified classification of meteorites is shown in Box 10.3.

PRIMITIVE METEORITES

The commonest meteorites are the **chondrites**, so called because most of them contain **chondrules** (millimetre-sized spheroidal assemblages of crystals and glass – MacKenzie *et al.*, 1982, plate 169). Chondrules are considered to be solidified droplets of melt formed by melting of dust in the early solar nebula during an early phase of heating. Nevertheless one can show that the mineral grains in the silicate–metal–sulphide matrix surrounding the chondrules have often not achieved chemical equilibrium, with each other or with the chondrule minerals, suggesting that temperatures during and after accretion of some chondrites have remained fairly low. **Carbonaceous chondrites** contain a complex, tarry organic component and various hydrous silicates of limited thermal stability.

Because they seem to have suffered the least thermal and chemical processing during the development of the solar system, the carbonaceous chondrites are said to be chemically **primitive**; they, particularly one group known as CI chondrites, are regarded as relics of the primordial solid matter of the solar system, and from these meteorites can be determined its overall composition, except for the most volatile elements.

DIFFERENTIATED METEORITES

Meteorites other than chondrites are products of the segregation of metal from silicate (forming 'irons' and achondrites respectively), and they are called **differentiated meteorites** (Box 10.3). It is generally assumed that differentiation is a consequence of incorporation into small planetary bodies, perhaps a few hundred kilometres across (Box 3.6, p. 72), in which high temperature facilitated gravitational segregation of the two phases, as in the case of our own planet.

Differentiated meteorites are poor representatives of primordial matter, but help in understanding the development of the planets.

10.2.3 Dark matter

In the foregoing we have assumed that all matter consists of known chemical elements – that is to say, consists ultimately of protons,

Box 10.3
Types of meteorite, shown in proportion to percentage of total falls[1]

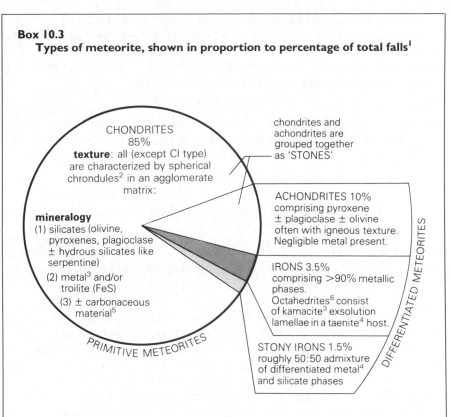

[1] 'Falls' are meteorite samples that have been seen to fall rather than just being discovered on the ground ('finds'). Relative numbers of 'falls' give the best estimate of the relative abundance of the different meteorite types.

[2] Chondrules are millimetre-sized spheroidal clusters of crystals and glassy material, thought to have originated as droplets of melt. The crystals (e.g. feathery pyroxene) show evidence of having been rapidly chilled (MacKenzie *et al.*, 1982, plate 169; Hutchison, 1983, plate 3).

[3] Kamacite = $Fe_{95}Ni_5$.

[4] Taenite $\sim Fe_{60}Ni_{40}$ (Box 3.6, p. 72).

[5] The carbonaceous component characteristic of the **carbonaceous chondrites** is a complex tarry mixture of abiogenic organic compounds.

[6] So called because the kamacite plates, owing to the cubic symmetry of taenite, have four orientations parallel to the faces of a regular octahedron. This distinctive texture, most apparent in a flat surface that has been etched with acid, is called Widmanstätten structure (Hutchison, 1983, Figures 4.11 and 4.12).

neutrons and electrons. As baryons (protons and neutrons) make up all but a tiny fraction of the mass of such matter, it is referred to as **baryonic matter**. It reveals its existence in distant parts of the universe by emitting or absorbing light, and thus its abundance is readily estimated from Earth by astronomical observation.

From such measurements astronomers are able today to estimate within quite narrow limits the overall density of baryonic matter in the universe as a whole, at least that part of it that is luminous and measurable. Big Bang cosmology defines three ranges within which the overall density of matter in the universe might fall:

(a)   A low-density universe whose gravitational self-attaction is too weak to reverse the cosmological expansion that began with the Big Bang.

(b)   A high-density universe in which gravitational forces are sufficient to halt and eventually reverse the expansion, leading to universal contraction and ultimately the end of the known universe in what has been called the 'Big Crunch'.

(c)   A critical density in between these two domains, just sufficient to halt the expansion at some infinitely distant time in the future, but not to reverse it.

Current cosmological orthodoxy holds that the density of matter in the universe should be exactly equal to the critical density. Astronomical estimates of the density of visible matter, however, amount to no more than 10% of the critical value. The discrepancy between theory and observation led to an intense debate in the late 1980s and early 1990s as to the nature of the 'missing' matter. Is it in the form of dispersed masses of baryonic matter that are simply too cool to emit detectable radiation ('dark' baryonic matter), or do we need to look for more exotic explanations involving matter of a totally novel kind lurking in and around the galaxies?

Though these are profoundly important questions for the cosmologist, they fall well outside the scope of this book. Dark matter, whatever its nature, seemingly plays no part in the formation of planets such as ours. It is nonetheless important to acknowledge that, when we generalize grandly about the composition of matter in the universe, we are referring solely to the 'visible' baryonic component that may account for as little as 10% of the mass of the universe as a whole.

10.3   The composite abundance curve

The two sources of information outlined above, solar spectra and analyses of primitive meteorites, allow us to build a composite picture of element abundance in the solar system as a whole. Gaseous elements – hydrogen, the inert gases and so on – can of course be determined only from solar measurements; for other elements, like boron, spectral measurements are difficult or impossible and reliance on meteorite data is the only feasible course. Fortunately the abundance of most elements can be determined by both methods. Since the two approaches involve different assumptions and employ different instrumental techniques, it is reassuring to find a good correlation between them (Figure 10.1).

The composite abundance data so obtained have been plotted against atomic number (Z) in Figure 10.2. Although compiled specifically for the solar system, the main features of this 'abundance curve' are common to practically all stars and luminous nebulae:

(a) Hydrogen and helium are several orders of magnitude more abundant than any other element. In atomic terms, helium has one-tenth of the abundance of hydrogen and together they comprise 98% of the solar system.

(b) Progressing to higher atomic numbers leads to an overall decrease in abundance, making the heaviest nuclei among the least abundant.

(c) The elements lithium, beryllium and boron are sharply depleted compared with the other light elements. (In the case of Li this depletion is much more marked for the Sun than for CI chondrites. Figure 10.1.)

(d) Elements having even atomic numbers are on average about ten times more abundant than elements with similar but odd atomic numbers. This effect, which is apparent in terrestrial rocks as

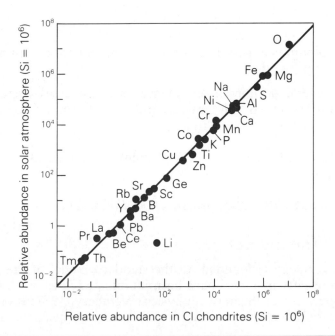

Figure 10.1 The correlation between element abundances in the Sun and in CI carbonaceous chondrites. Abundance is expressed as the number of atoms of each element per 10^6 atoms of silicon.

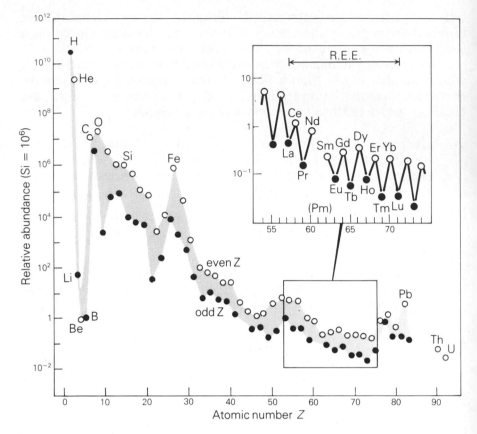

Figure 10.2 The composite solar-system abundance curve. The vertical axis shows on a logarithmic scale the number of atoms of each element per 10^6 atoms of silicon. The inset shows abundances of the rare earth elements ('REE', La through Lu) and some neighbouring elements. The element promethium (Pm) has no stable isotopes and is not found in meteorites (Exercise 1 at the end of this chapter).

○ Elements with Z even
● Elements with Z odd

The stippled band indicates the abundance difference between even-Z and odd-Z nuclides.

well, produces a 'sawtooth' profile if adjacent atomic numbers are joined up (see inset).

(e) The general decline in abundance as Z increases (item (b)) is interrupted by a sizable peak around $Z = 26$, comprising elements in the neighbourhood of iron.

These features are clues to how the elements were formed.

10.4   Cosmic element production

10.4.1   The Big Bang

Modern cosmology is founded on the standard model of the Hot Big Bang, whose early moments have been illuminated in astonishing detail by the application of theoretical physics (Weinberg, 1977; Riordan and Schramm, 1991). Matter, in the form of protons, neutrons and electrons, began to take form in the expanding primal fireball after only the first second of time, at a temperature of about 10 billion degrees (10^{10} K). Theory predicts that 25% of the baryons were neutrons, the remainder protons. After the first few minutes, neutrons combined with protons to form a sprinkling of light nuclei such as ^{2}H (deuterium), ^{3}He and ^{4}He. The temperature (10^9 K) was still far too hot at this stage for nuclei to capture electrons; matter in the universe had to wait 100 000 years or more before the temperature had fallen low enough ($\sim 5 \times 10^3$ K) for neutral atoms to form.

One of the resounding triumphs of 'Big Bang cosmology' over the last three decades (a story recounted in the fascinating book by Riordan and Schramm, 1991) is how closely the light nuclide abundances that are predicted by theoretical modelling – 75% ^{1}H, 100 ppm ^{2}H, 20 ppm ^{3}He, 25% ^{4}He and 0.5 ppb ^{7}Li by mass – match the mean abundances actually measured in the cosmos today. From this we can infer (1) that the Big Bang hypothesis is correct in essential details, and (2) that the present cosmic inventory of these lightest nuclides is a relic of the Big Bang and not, as scientists once believed in relation to He and Li, the product of more recent processes.

10.4.2 Stars

These light nuclides provided the feedstock for the manufacture of all the heavier chemical elements that make up the Earth and the other traces of solid matter in the cosmos. This manufacturing process has been going on throughout the life of the Universe. It has evidently not been very 'efficient' as ^{1}H and ^{4}He still make up 98% of the mass of the known (baryonic) universe, as noted in the list above. How have these heavier elements been formed?

The principal process for generating the elements up to iron is nuclear fusion (Box 10.2). Fusion can only occur if two conditions are both satisfied:

(a) a high density of matter to raise the probability of nuclear collisions (which in interstellar space is vanishingly small), and
(b) a high temperature (at least 10^7 K) to ensure that nuclei will collide with sufficient kinetic energy to overcome their mutual electrostatic repulsion; nuclei need to approach to within 10^{-14} m before the Strong Nuclear Interaction begins to operate.

The interior of a star furnishes both of these requirements, and the abundances of heavier elements that we see today are regarded as the cumulative product of nucleosynthesis that has taken place in many generations of stars.

We can visualize stellar nucleosynthesis as a long series of consecutive steps, like an industrial assembly line. Not every stellar factory, however, possesses the full assembly line. Fusion reactions in stars take place in a series of stages, hand in hand with the thermal evolution of the star, and how far the process may go depends, as we shall see, on the mass of the star.

Hydrogen is consumed to form helium early in the development of a star (see table). When hydrogen in the centre of the star is almost used up, the star raises its temperature by gravitational contraction to a level sufficient to allow helium nuclei to combine to form carbon and oxygen. Similarly helium must more or less run out in the core of the star before carbon and oxygen can be transformed to the heavier elements leading to silicon.

Stage	Maximum temperature	Range of nuclei produced
1	10^7 K	H $\rightarrow$ He
2	10^8 K	He $\rightarrow$ C, O, etc.
3	5×10^8 K	C, O $\rightarrow$ Si
4	5×10^9 K	Si $\rightarrow$ Fe

The maximum temperature a star can achieve during normal evolution is related to its mass. A star of the Sun's mass ($M_\odot$) is capable only of stages 1 and 2. A star probably needs to have mass exceeding $30 M_\odot$ before all fusion reactions leading to iron become possible (Tayler, 1975). Many stars fall below this range, and therefore contribute only to the abundance of the lighter elements. The general fall-off in abundance towards heavier nuclides (item (b) on p. 241) reflects the relatively small number of stars capable of generating the heaviest elements.

Fusion reactions can generate most but not all of the stable nuclides between hydrogen and iron. As ^{8}Be is a very unstable nucleus, the main fusion reactions evidently proceed directly from ^{4}He to ^{12}C, largely bypassing the elements Li, Be and B. The small amounts of Be and B found in nature ((item (c) on p. 241) seem to have been produced by the breakdown of heavier nuclei (^{12}C, ^{16}O) under cosmic-ray bombardment, a process called **spallation**.

The balance of nuclear forces gives nuclei in the iron mass range the greatest stability (Box 10.2). Massive stars having sufficiently high temperatures can produce these nuclides very efficiently. Beyond this point however further fusion is impeded, because the temperatures required to overcome the electrostatic repulsion between such highly charged nuclei exceed those of the hottest stars. The synthesis of heavier nuclides proceeds almost entirely by the addition of **neutrons**, neutral nuclear particles that are not repelled by the host nucleus. Many reactions in stars produce neutrons, particularly in the latter stages of stellar evolution. The nuclides to the right of iron in Figure 10.2 are produced by cumulative **neutron capture**. Neutrons are absorbed by a nucleus, increasing the N value until an unstable, neutron-rich isotope has been produced, which transmutes by β-decay into an isotope of the next element (Z increases by one, N falls by one; Box 10.1). The repeated operation of this process can produce all of the heavier nuclides, given a sufficiently high neutron flux (Henderson, 1982 and Tayler, 1975 for details). The abundance peak around iron in Figure 10.2 (item (e) on p. 242) suggests that this process consumes iron-group nuclei more slowly than fusion reactions produce them.

Why are nuclides that have even values of Z – or N, for that matter – more abundant than those with odd values (item (d) on p. 241)? Protons and neutrons reside in orbitals inside the nucleus just as electrons do outside it. According to nuclear wave mechanics, filled orbitals containing two protons or two neutrons are more stable than half-filled ones. This additional stability is expressed in the form of a more compact nucleus. This reduces the nuclear 'cross-section' upon which the probability of collision depends, thereby depressing the rate of the fusion or neutron-capture reactions that consume the nuclide, and allowing its abundance to build up. Even values of Z and N account for the greatest number of stable nuclides, lending the nuclide chart a 'staircase' appearance in which even values of Z form the treads and even values of N form the steps (Box 10.1).

10.4.3   Supernovae

When a smaller star ($<M_\circ$) reaches the end of its life, it can progress into a 'white dwarf' phase and quietly fade away. But theory suggests that massive stars ($>2M_\circ$) follow a different path, leading to a catastrophic collapse of the core which causes a colossal stellar explosion (Bethe and Brown, 1985). Such **supernovae** are characterized, for a brief period, by energy output of staggering intensity. The luminosity from a single exploding star can rise to levels typical of a whole galaxy ($\sim 10^{11}$ stars), lasting for a few Earth days or weeks. The huge quantities of energy transferred to the zones of the star immediately surrounding the core cause a large proportion of the star's mass to be expelled at high velocity ($\sim 10^7 \, \mathrm{m\,s^{-1}}$). The expanding Crab Nebula is thought to be the remnant of a supernova observed by Chinese astronomers in AD 1054. A supernova was actually observed in the Large Magellanic Cloud on 23 February 1987.

Supernovae contribute to nucleosynthesis in two important ways:

(a)   The neutron flux during a supernova is exceedingly high, prompting a burst of very rapid neutron-capture reactions leading to U and Th, and even beyond (to heavy unstable nuclides such as plutonium, Pu).

(b)   The products of nucleosynthesis confined within a star's interior are flung back into the interstellar medium, to be incorporated into new generations of stars. Present element abundances (Figure 10.2) reflect recycling of matter through successive generations of

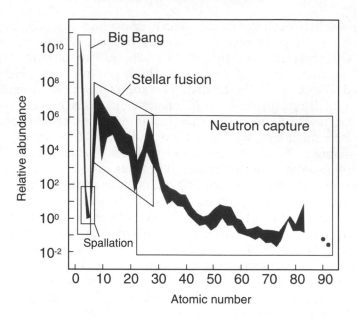

Figure 10.3 Solar-system abundance curve showing domains of various nucleosynthetic processes.

stars, each one adding its own contribution to the overall accumulation of heavy elements in the universe. As Hutchison (1983) points out, 'we, each one of us, have part of a star inside us'.

Figure 10.3 summarizes the contribution of these different processes to the current inventory of chemical elements in the universe.

10.5   Elements in the solar system

10.5.1   Cosmochemical classification

Differentiated meteorites contain three broad categories of solid material: silicate, metal and sulphide. Analysis of these phases shows that most elements have a greater affinity with one of them than with the others. Magnesium, for example, is overwhelmingly segregated into silicate phases, whereas copper is often concentrated in sulphides. The Norwegian geochemist V. M. Goldschmidt introduced the following subdivision:

(a)   lithophile elements: those concentrated into the silicate phase (from the Greek *lithos*, meaning 'stone');

(b)   siderophile elements: those preferring the metal phase (from the Greek *sideros*, meaning 'iron');

(c)   chalcophile elements: those like copper, which concentrate in the sulphide phase (from the Greek *chalcos*, meaning 'copper');

(d)   atmophile elements: gaseous elements (from the Greek *atmos*, meaning 'steam' or 'vapour').

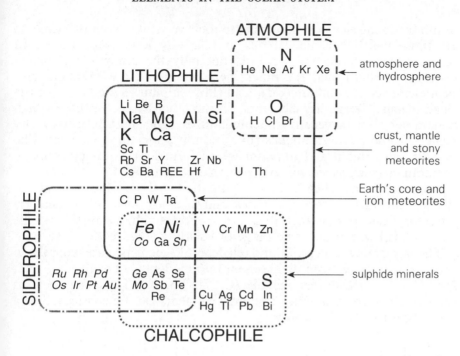

Figure 10.4 Element affinities in the Earth and meteorites. Areas of overlap show elements common to two or more phases. Larger lettering indicates a major element. Elements found principally in the iron phase are show italicized. (REE – Figure 10.2.)

How the elements are divided between these categories is shown in Figure 10.4. Such a compilation involves compromise, and one author's version will differ slightly from another's. Among the metals there is a significant correlation with electronegativity (Box 9.8): the lithophile metals have electronegativities below 1.7, most chalcophile metals have electronegativities between 1.8 and 2.2, and the siderophile metals are those with electronegativities of 2.2 and above. Goldschmidt's concept is very useful in understanding in what form elements occur in solar-system matter, in ore deposits or for that matter in a smelter. For example, the siderophile character of iridium (Ir) means that nearly all of the Earth's Ir inventory is locked away in the core (the same incidentally being true of gold) and the concentration in crustal rocks is extremely low (Figure 9.1). Consequently most of the iridium one detects on the Earth's surface, in deep-sea sediments for instance, has been introduced as a constituent of meteoritic dust; some iron meteorites contain as much as 20 ppm Ir, 20 000 times higher than average levels in crustal rocks. This provides a means of estimating the annual influx of iron meteorites to the Earth's surface.

Some elements exhibit more than one affinity. For example, oxygen is both lithophile – as a major constituent of silicates – and atmophile (as O_2 and H_2O).

In considering the development of the solar system, it is also useful to subdivide the elements according to their volatility. **Volatile** elements are those which become gaseous at relatively low temperatures. In cosmochemical terms they include not only the atmophile elements hydrogen, helium (and the other inert gases) and nitrogen, but also such elements as cadmium (Cd), lead (Pb), sulphur (S) and most of the alkali metals. **Refractory** elements, on the other hand, are those which remain solid up to very high temperatures. The most refractory elements are the platinum metals (like iridium), and those elements like calcium, aluminium and titanium which form highly refractory oxide compounds (such as the mineral perovskite, $CaTiO_3$).

Magnesium and silicon, the elements that make up the bulk of the silicate minerals in the meteorites and planets, form a 'moderately refractory' category between these two extremes. The major siderophile elements fall within the same range of volatility (Figure 10.6).

The remaining lithophile and chalcophile elements are volatile in varying degrees. We can divide them into moderately volatile (e.g. Na, Mn, Cu, F, S) and very volatile (C, Cl, Pb, Cd, Hg) categories, as shown in Figure 10.6. The atmophile elements can be considered as a third, 'most volatile' category.

10.5.2   Element fractionation in the solar system

It has long been known that the planets differ considerably in composition. During the development of the solar system, the elements have been chemically sorted or **fractionated**.

Because metal, silicate and gas phases themselves differ in density, planetologists are able to estimate the proportions of these materials present in the planets (whose densities can be determined from astronomical measurements). As one can see in Figure 10.5, the planets differ considerably in their make-up. The small inner planets Mercury, Venus, Earth and Mars – the **terrestrial planets** – have high densities characteristic of mixtures of metal and silicates, though in varying proportions. The atmospheres make up only a trivial proportion of planetary mass.

With the exception of Pluto, the remaining planets – the **major planets** – have masses one or more orders greater than the Earth's. Their low densities ($0.69\,kg\,dm^{-3}$ for Saturn to 1.64 for Neptune) indicate compositions much closer to the solar-system average, in which atmophile elements predominate. The largest planet, Jupiter, consists almost entirely of hydrogen and helium, although with a small rock and ice core of 10–20 times the Earth's mass.

More is known about the constitution of the Earth and Moon than the other planets, and geochemists have been able to assemble by indirect means (Taylor, 1982) quite detailed models of the overall composition of these two bodies. They are illustrated in Figure 10.6. The Earth and Moon are depleted not just in atmophile elements

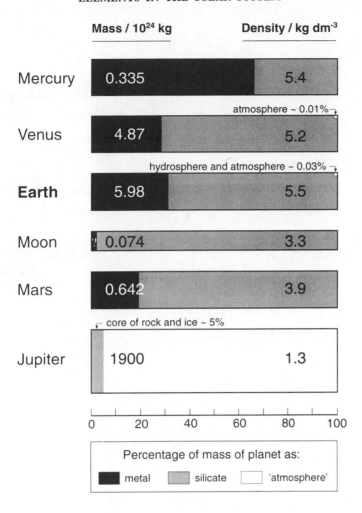

Figure 10.5 Proportions of metal, silicate and 'atmosphere' in the terrestrial planets and Jupiter, from astronomical data. It is not certain on geophysical grounds whether the Moon has a metallic core, but the existence of a small one is geochemically probable.

(Figure 10.5), but in the other volatile elements too. Note that this depletion and the slight enrichment of the most refractory elements are more prominent in the Moon than the Earth.

10.5.3 Evolution of the solar system

There is a general (but not universal) consensus that the Sun and the solar system developed together by gravitational condensation in a large cloud of interstellar gas and dust about 4.6 billion years ago. The overall composition of this **nebula** must have been close to the present solar-system average shown in Figure 10.2. The heavy elements in the cloud had presumably accumulated from a number of separate episodes of nucleosynthesis.

The potential energy released by gravitational collapse would have

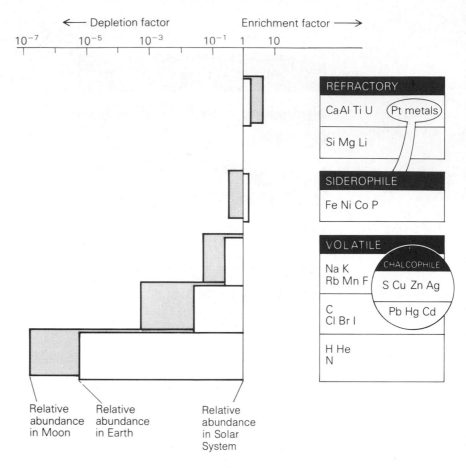

Figure 10.6 Element fractionation in the Earth (blank bars) and the Moon (stippled bars). The length of the bars indicates the approximate degree of depletion (left) or enrichment (right) of each group of elements compared to solar-system average abundances – see scale at top. Because Si is the reference element for expressing both terrestrial/lunar and solar-system abundances (Figure 10.2), the Si group of elements registers neither enrichment nor depletion in this diagram.

appeared in the form of heat in the denser parts toward which the collapse was directed, leading eventually to temperatures high enough for thermonuclear fusion to begin. Our interest is mainly in the peripheral parts of the solar nebula, from which the planets have since developed.

Given the huge variations in the abundances of the volatile elements in the solar system, it is natural to postulate that the solar nebula was initially very hot. The **equilibrium condensation theory** regards the solid constituents of the solar system as condensates from a cooling, entirely gaseous nebula whose initial temperature was in the region of 1500 °C. The elements would condense into solids in a predictable 'condensation sequence' which, given one or two assumptions about the density and composition of the gas, can be worked out thermodynamically. First to condense would be the most refractory elements – the platinum metals and oxides of Ca, Al, Ti, and so on – appearing as solids at about 1600 K. Other elements would follow at lower temperatures, broadly in the descending order shown in Figure 10.6. The planets would differ in their content of volatile elements according to their distance from the Sun, or to the stage in this cooling sequence at which they accreted to planetary size. Those accreting early would fail to incorporate volatile elements, not yet available in solid form; whereas accretion at a later stage or in a cooler part of the nebula would lead to assimilation of the lower-temperature condensates too, producing material of more primitive composition like carbonaceous chondrites.

Oxygen isotope studies suggest, however, that the early solar system was not as homogeneous as one would expect of a wholly gaseous nebula. Moreover, astronomical observations of nebulae in which stars are believed to be forming suggest that they are relatively cold (20–30 K). Some geochemists now prefer a theory of **cool accretion** from a nebula consisting of both gas and dust. Astronomers have suggested that the Sun, early in its development, may have passed through a 'T-Tauri' phase (named after a star showing this stage of evolution) in which the intensity of the **solar wind** – the outward flux of protons radiating from the Sun – was much greater than now. Gaseous and volatile elements could have been swept out from the inner parts of the solar nebula under the bombardment of this intensified solar wind, leaving the inner planets to accrete from a volatile-depleted residue.

How did a solar nebula consisting of finely dispersed dust and gas accrete into the large planets that exist today? The current view is that the nebula became progressivly more 'lumpy' as small particles, on colliding, clumped together into larger ones. Collisions would have been frequent, breaking up some bodies but adding to others. As time went by the particle-size distribution evolved through metre-scale and kilometre-scale bodies to 100-km scale planetesimals – a sample of which is preserved today in the asteroid belt. By sweeping up smaller bodies in their path, such bodies would grow larger and fewer in number, eventually coalescing into the present planets.

This main accretionary stage in the formation of the planets probably lasted only 100 million years. Yet it is clear from the density of cratering on the Moon – and from radiometric dating of younger, less cratered lunar basalt terrains – that intensive bombardment of the terrestrial planets by smaller planetesimals continued until about 3800 Ma ago, 700 million years after the planets had originally been formed. Some of the impacts early in this period were evidently huge. The anomalously large core of Mercury has been attributed to a giant impact that eliminated a large proportion of its original silicate mantle. The Earth is also considered to have suffered a catastrophic impact, in which the mantle of the colliding planetesimal – whose mass was about 14% of the Earth's – was ejected into orbit to form our present Moon (thereby accounting for its low density in relation to other inner-solar system bodies) while its core was captured by the Earth. Such an event would explain the extreme volatile depletion of the Moon (Figure 10.5) if the impactor mantle were initially vaporized by the collision.

The solar system's early history is described in more detail by S. R. Taylor and D. L. Anderson in Brown *et al.* (1992).

10.6 Chemical evolution of the Earth

10.6.1 The core

If the Earth was formed by the aggregation of large planetesimals, the energy of their collisions would certainly have been sufficient for the early Earth to have been entirely molten, leading to immediate gravitational separation of dense metal to form the core. The impact event proposed for generating the Moon would have caused a second phase of extensive melting and an additional contribution to the core.

Many physical properties of the core are consistent with a major element composition similar to the Fe–Ni alloy found in iron meteorites, but the velocity of compressional seismic waves through the core indicates a lower density than expected for Fe–Ni under the appropriate load pressure, and it follows that a significant proportion (~10%) of some less-dense element must also be present in the core. Though either sulphur or oxygen would fit the bill on density grounds, current opinion favours sulphur.

The gravitational accumulation of dense metal into the centre of the molten Earth has two important implications: large amounts of gravitational potential energy would be released, sustaining the molten state of the overlying mantle, and siderophile elements (Figure 10.4) would be efficiently scavenged from the mantle into the core.

10.6.2 The mantle

The silicate material surrounding the core, comprising 70% of the Earth's mass (Figure 10.5), has now differentiated into the mantle and crust, as a result of continual igneous activity throughout the Earth's history. When a partial melt (Box 2.4) develops in equilibrium with solid rock, elements are fractionated in two overlapping ways.

(a) The lower-melting major components of the rock (Fe, Al, Na, Si) enter the melt preferentially, leaving the residual solids enriched in refractory (Mg-rich) end-members (Box 2.4).

(b) Crystals tend to dump into the melt certain trace elements which are difficult to accommodate. The ions of these **incompatible elements** (Box 9.1) are more easily accommodated in the open, disordered structure of a melt than in a crystal lattice.

Extraction of magma from the mantle progressively displaces these elements from the mantle into the crust. Crustal rocks (e.g. basalt)

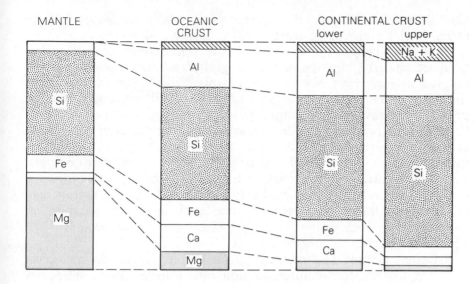

Figure 10.7 Average compositions (oxide percentages) of the Earth's mantle and crust.

consist of lower-melting mineral assemblages (Figure 10.7) and are enriched in the incompatible elements compared with the mantle.

Parts of the mantle have therefore become depleted in these elements. It has been estimated that 25–50% of the mantle's original budget of highly incompatible elements (K, Rb, U) now resides in the continental crust. Supposing these elements were uniformly dispersed in the primordial mantle, it seems that at least one-third of its volume has been tapped for igneous magma during the course of geological time. Whether the mantle was ever homogenous is debatable (though it seems likely if at one time the mantle was completely molten), but its present inhomogeneity is beyond doubt. The varying geochemistry of recent volcanic rocks points to a range of chemically distinct source regions in the mantle. Most mid-ocean ridge basalts ('MORBs'), for example, come from mantle domains depleted in the incompatible elements – an inheritance, most geochemists believe, from widespread melt extraction during the course of Earth history.

10.6.3 The crust

The Earth's crust falls into two broad types. The crust of the ocean basins, being the product of partial melting of mantle peridotite, has a relatively high Mg content (though much lower than the mantle itself) and low Si (less than 50%). It has a short lifetime: its passage from ocean ridge to subduction zone, where it is delivered back into the mantle, takes less than 200 million years (the age of the oldest known oceanic crust). During this time it interacts chemically with ocean water and acquires a blanket of sediment. Parts of this modified crustal package are recycled as constituents of the island-arc volcanics and

253

Cordilleran plutonic rocks that appear above subduction zones, but the extent of this recycling of crust is still not fully understood.

Unlike the ephemeral ocean floor, the **sialic** continental crust has been accumulating throughout known geological time, though probably not at a uniform rate. None of the Earth's earliest crust has survived: the oldest recognizable remnants are about 4.0×10^9 years old (although greater ages have been obtained from detrital mineral grains). Continental crust is being extended today by the lateral accretion of island arcs on to continental margins, and by deep-seated igneous intrusions into their roots. Different mechanisms may have operated in the past, for example during the huge increase in the volume of the continental crust that seems to have occurred between 3.0 and 2.0 billion years ago, at the close of the Archaean era. Average continental crust is equivalent to andesite in composition, having a SiO_2 content of around 57%, significantly higher than oceanic crust.

Repeated melting and metamorphism within the continental crust have led to its internal differentiation into a lower, more refractory crust depleted in incompatible elements; and an upper crustal layer, roughly 10 km thick, which is enriched in Na, K and Si (Figure 10.7). Most of the heat flow we measure in continental areas originates in this top 10 km, into which almost the whole crustal inventory of the radioactive incompatible elements K, U and Th has been concentrated.

It is notable that the other terrestrial planets have dominantly basaltic crust. Why has none developed andesitic or 'granitic' crust resembling the continents on Earth? This unique feature of the Earth probably stems from another, the existence on the surface of liquid water. The igneous minerals of the basaltic ocean crust react with sea water to form hydrous secondary minerals. Subduction of altered oceanic crust transports this bound water deep into subduction zones where the hydrous minerals dehydrate (cf. Figure 2.3), thereby releasing water into the overlying mantle wedge. In the presence of water vapour, mantle peridotite melts at lower temperatures and – as experiments have shown – produces melts that are more SiO_2-rich (andesitic) than when melted 'dry'. This explains the dominance of andesite in many island arcs, and provides the precursor for the formation of granitic upper continental crust.

10.6.4 The atmosphere

From the point of view of a living creature, the most important aspect of Earth evolution is the development of the present atmosphere and hydrosphere. It seems certain that the primordial gases of the solar nebula had been swept away from the terrestrial planets prior to the Earth's formation, and would in any case have been lost during the molten inferno accompanying accretion. The earliest terrestrial atmosphere was therefore made up of residual atmophile elements still trapped in the Earth's interior and swept out ('degassed') by volcanic activity, and almost certainly was similar in composition to the atmosphere of Venus today: mainly carbon dioxide, with small amounts of N_2, SO_2, H_2O and Ar. There is little doubt that it was entirely devoid of free oxygen, and the earliest life forms must have been able to survive in this **anoxic** environment. In one sense the comparison with Venus is false, however; there is evidence for the existence of liquid water on the surface of the Earth since at least 3.8×10^9 years ago, so the temperature must have been much lower than at the surface of Venus today (490 °C).

The ultimate origin of life remains shrouded in mystery, but among the earliest organisms were some, probably not unlike today's cyanobacteria ('blue-green' algae), that produced oxygen in the oceans as a by-product of the photosynthesis of carbohydrate from carbon dioxide and water (Equation 9.3, p. 214). This biogenic oxygen could not immediately accumulate in the atmosphere, however, because the oceans would already have accumulated through weathering a large inventory of reduced solutes, notably Fe^{2+}. Until completely oxidized, this 'oxygen sink' would have mopped up oxygen as it was evolved, preventing it from reaching the atmosphere. There is still debate about when the elimination of the oxygen sink was achieved. The orthodox view places the critical transition at about 2000 million years ago, when a gross change occurred in the dominant style of iron deposition in sediments. In the sediments of the preceding Archaean eon, iron occurs chiefly in banded iron formation, the fine iron-rich layers of which were precipitated from sea water through the oxidation of dissolved Fe^{2+}. Each layer may represent a single short-lived bloom of oxygenic bacteria. In later sediments iron occurs more commonly as rusty diagenetic coatings on detrital grains in reddish sandstones or shales ('red beds'), suggesting that iron released by weathering was being oxidized subaerially long before it reached the sea, presumably owing to abundant free oxygen present in the atmosphere by that stage.

The manner in which life adapted to the prevalence of oxidizing conditions at the Earth's surface, putting oxygen to good use to generate energy, is a fascinating story that can be followed up in the book by Nisbet (1991). The earliest life-forms evidently developed on an abundant supply of organic nutrients that could exist stably in he oxygen-free primordial atmosphere. Such organisms could not develop in the present atmosphere, in which oxidation would rapidly destroy their simple molecular foodstuffs, any more than we could survive in oxygen-free conditions. Life, by introducing free oxygen into the atmosphere and maintaining its level for at least 2 billion years (the residence time of oxygen in the atmosphere is only a few thousand years), has burned the environmental boat by means of which it came into being.

Yet life has also made the Earth into the tolerable place to live in that it currently is. All of the oxygen in the atmosphere has been manufactured by photosynthetic organisms from carbon dioxide, and such organisms have thereby turned down the heat in the Earth 'greenhouse' to a much lower level than operates on Venus, whose 'greenhouse effect' maintains the surface temperature at a searing 490 °C. This mechanism for removing CO_2 from the terrestrial atmosphere is essentially a reversible equilibrium, balancing oxidized carbon in the air against reduced carbon in the biosphere, though some of the reduced carbon also becomes fixed in the lithosphere in the form of fossil fuel. Equally important from the 'greenhouse' point of view, however, is the myriad of carbonate-secreting organisms that over the eons of geological time have fixed atmospheric carbon in the form of limestone (the product of accumulation of calcareous biogenic debris). Together these carbon reservoirs in the crust account for the very marked difference in atmospheric CO_2 between Earth and Venus (0.03% versus 96.5%) and for the temperate climate that we enjoy on Earth. Let us take care to keep it so.

10.7 Bibliography

Barrow, J. D. and Silk, J. (1983) *The left hand of creation*, Unwin Paperbacks, London.

Bethe, H. and Brown, G. (1985) How a supernova explodes. *Scientific American*, **252**(5), 40–8.

Brown, G. C., Hawkesworth, C. J. and Wilson, R. C. L. (1992) *Understanding the Earth – a new synthesis*, Cambridge University Press, Cambridge, Chapters 1–3.

Cloud, P. (1978) *Cosmos, Earth and man*, Yale University Press, New Haven, Conn.

Henderson, P. (1982) *Inorganic geochemistry*, Pergamon, Oxford.

Hutchison, R. (1983) *The search for our beginning*, Clarendon Press, Oxford.

MacKenzie, W. S., Donaldson, C. H. and Guilford, C. (1982) *Atlas of igneous rocks and their textures*, Longman, Harlow.

Nisbet, E. G. (1991) *Living Earth – a short history of life and its home*, Chapman & Hall, London.

Open University (1994) *The planets*, Open University, Milton Keynes.

Riordan, M. and Schramm, D. M. (1991) *The shadows of creation – dark matter and the structure of the universe*, W. H. Freeman, New York.

Tayler, R. J. (1975) *The origin of the chemical elements*, 2nd edn, Wykeham, London.

Taylor, S. R. (1982) *Planetary science: a lunar perspective*, Lunar and Planetary Institute, Houston.

Taylor, S. R. and McLennan, S. M. (1985) *The continental crust: its composition and evolution*, Blackwell, Oxford.

Weinberg, S. (1977) *The first three minutes*, Fontana.

10.8 Exercises

1 Examine the values of Z and N (nuclide chart, Box 10.1) for which there are no naturally occurring nuclides. What do they have in common? Why?

2 The decay of the isotope ^{26}Al (half-life 0.7 million years) is thought to have been an important source of heat during the early history of the solar system. Calculate (a) the decay constant of ^{26}Al, and (b) the time required for the rate of heat production to fall to one-hundredth of its initial value.

Answers to exercises

Chapter 2

1   Point X Phases present are calcite + quartz + CO_2 gas.
    $\phi = 3$. $C = 3$ (CaO, SiO_2, CO_2).
    $3 + F = 3 + 2 \rightarrow F = 2$.
    Point Y ϕ = calcite + quartz + wollastonite + $CO_2 = 4 \rightarrow F = 1$.
    Temperature and P_{CO_2} can vary independently at point X without changing the equilibrium assemblage.

2   The lower density of ice indicates that at $0\,°C$ $V_{ice} > V_{water}$. For the reaction

$$H_2O \rightleftharpoons H_2O$$
$$\text{ice} \qquad \text{water}$$

ΔS = +ve (always true for melting).
ΔV = −ve.   Therefore $dP/dt = \Delta S/\Delta V$ = −ve.
The negative slope of the melting curve indicates that the melting temperature falls as pressure is increased.

3      $\Delta S = 202.7 + (2 \times 82.0) - 241.4 - 41.5 = 83.8\,J\,K^{-1}\,mol^{-1}$
      $\Delta V = 32.6 \times 10^{-6}\,m^3\,mol^{-1}$
    $\Delta S/\Delta V = 2.57 \times 10^6\,J\,K^-\,m^{-3}$
          $= 2.57 \times 10^6\,Pa\,K^{-1}$
We know one point on the reaction boundary ($10^5\,Pa$ at $520\,°C$), At $520 + 300\,°C$ the pressure on the boundary will be $300 \times 2.57 \times 10^6\,Pa = 7710 \times 10^8\,Pa$. As all phases are anhydrous, a straight reaction boundary is expected between these two points. The volume of the grossular + quartz assemblage is less than the anorthite + 2(wollastonite) assemblage, and therefore it is found on the high-pressure side.

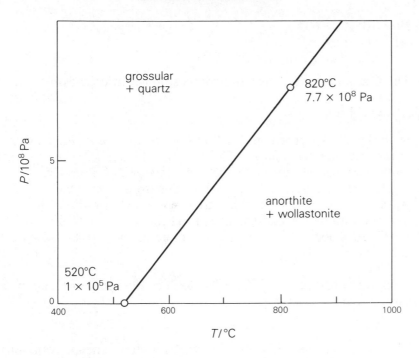

4 (a) 1400°: liquidus to vertical dashed line is 3.5 mm.
Solidus to vertical line is 27 mm.
Liquid/crystals = 27/3.5 = 7.7

$$or \text{ Percent liquid} = \frac{27 \times 100}{(3.5 + 27)} = 89\%$$

Liquid composition = An_{35}. Crystals An_{72}.
  (b) Liquid/crystals = 0.60. Percent melt = 37%.
Liquid An_{15}. Crystals An_{53}.
  (c) Liquid/crystals = 0. Liquid An_7. Crystals An_{40}.

5 Required diopside–plagioclase mixture is d:
$$\text{Percent diopside in } d = \frac{24}{27.5 + 24} \times 100 = 46.6\%.$$

$$\text{Percent plagioclase } c = \frac{27.5}{27.5 + 24} \times 100 \times 53.4\%$$

Solid mixture a consists of plagioclase (average composition $f =$ An_{31}) and diopside. Lever rule gives 41% f and 59% diopside.
a lies on the boundary of the three-phase triangle melt–diopside–plagioclase at 1220 °C.
Equilibrium assemblage is melt b 56%, diopside 44% and plagioclase c 0%.

259

6 $20 \times 10^8 \, \text{Pa} \equiv 60 \, \text{km}$
   (a) $\text{Di}_{69} \, \text{Fo}_{07} \, \text{En}_{24}$
   (b) $\text{Di}_{69} \, \text{Fo}_{07} \, \text{En}_{24}$

The first melt to form has the composition E in each case.

Chapter 3

1 $N_0 = 18\,032$. Calculate $\ln(N_0/N)$ for each value of t. (As a check on arithmetic, the value for $t = 25$ is 0.09426.) $\text{Ln}(N_0/N)$ gives a linear plot against time, indicating that the reaction is first-order.
   After one half-life, $N/N_0 = 1/2$, therefore $\ln(N_0/N) = 0.6913$. Reading from the graph, this value is reached at $t = 190$ hours. If $n =$ the number of half-lives required for decay to $1/100$, $(1/2)^n = 1/100$, so $n \, \log(1/2) = \log(1/100)$, therefore $n = 6.6$ and $t_{1/100} = 1254$ hours.

2 Room temperature $= 25\,^\circ\text{C} = 298 \, \text{K}$. The doubling of reaction rate can be written:

$$k_{308} = 2k_{298}$$

The Arrhenius equation in log form gives simultaneous equations:
at 298 K: $\ln(k_{298}) = \ln A - E_a/(8.314 \times 298)$
at 308 K: $\ln(2k_{298}) = \ln A - E_a/(8.314 \times 308)$
Therefore $\ln A = \ln k_{298} + E_a/2478 = \ln 2k_{298} + E_a/2561$.
Rearranging, $\ln 2k_{298} - \ln k_{298} = \ln 2 = E_a(1/2478 - 1/2561)$.
$E_a = 0.6915/0.013 \times 10^{-3} = 52\,900 \, \text{J mol}^{-1} = 52.9 \, \text{kJ mol}^{-1}$

3 Calculate $\ln(1/\text{viscosity})$ for each temperature and plot against $1/T$. (E.g. for $T = 1325\,^\circ\text{C} = 1598 \, \text{K}$, $1/T = 0.000\,626 \, \text{K}^{-1}$ and $\ln(1/\eta) = -7.622$.) Slope of graph $= -34\,030 \, \text{K} = -E_a/R$. Thus $E_a = 283 \, \text{kJ mol}^{-1}$.

4 Half-life $= \ln 2/\lambda_{87_{\text{Rb}}} = 4.9 \times 10^{10}$ years.
   Therefore $\ln(N_0/N) = \lambda t = 1.42 \times 10^{-11} \times 4.6 \times 10^9$
   $\qquad\qquad\qquad\qquad = 0.0653$
   $\qquad\qquad \therefore N/N_0 = 94\%$.

Therefore percentage decayed $= 6\%$.

Chapter 4

1 (a) $\text{BaSO}_4 \rightleftharpoons \text{Ba}^{2+} + \text{SO}_4^{2-}$
   $\qquad\quad$ solid $\qquad\qquad$ solution

$$K_{\text{BaSO}_4} = a_{\text{Ba}^{2+}} \cdot a_{\text{SO}_4^{2-}} = \frac{m_{\text{Ba}^{2+}}}{m^\ominus} \cdot \frac{m_{\text{SO}_4^{2-}}}{m^\ominus} = 10^{-10}$$

When dissolved in pure water: $a_{Ba^{2+}} = a_{SO_4^{2-}} = 10^{-5}$
Therefore if solution is ideal, $m_{Ba^{2+}} = m_{SO_4^{2-}} = 10^{-5} m^{\ominus}$
$= 10^{-5}\,mol\,kg^{-1}$ $BaSO_4$ in saturated solution.
(b) In $CaSO_4$ solution, $a_{SO_4^{2-}} = a_{Ca^{2+}} = 10^{-3}$
If $x.m^{\ominus}\,mol\,kg^{-1}$ of $BaSO_4$ dissolve:
$K_{BaSO_4} = a_{Ba^{2+}} \cdot a_{SO_4^{2-}} = 10^{-10}$
$\qquad\quad = x(10^{-3} + x) \cong 10^{-3}x$ (since x^2 is very small)
Therefore $x.m_{\ominus} = 10^{-7}\,mol\,kg^{-1}$.

2 $CaF_2 \rightleftharpoons Ca^{2+} + 2F^-$
   solid   solution
$K_{CaF_2} = a_{Ca^{2+}} \cdot (a_{F^-})^2$
In pure water $a_{F^-} = 2a_{Ca^{2+}}$
Therefore $K_{CaF_2} = a_{Ca^{2+}} \cdot (2a_{Ca^{2+}})^2 = 4(a_{Ca^{2+}})^3 = 10^{-10.4} \cong 4 \times 10^{11}$
Therefore $a_{Ca^{2+}} = 0.000\,22$
$\qquad\qquad m_{Ca^{2+}} = 0.000\,22\,mol\,kg^{-1}$
Relative molecular mass of $CaF_2 = 40 + (2 \times 19) = 78$
Therefore $0.000\,22 \times 78 = 0.017\,g$ CaF_2 will dissolve in 1 kg water
at 25 °C.

3 $CO_2 + H_2O \rightleftharpoons H_2CO_3$
$K = a_{H_2CO_3}/(P_{CO_2})^{air} = 0.031$
$a_{H_2CO_3} = 0.031 \times 0.0003 = 0.000\,0093 = 10^{-5.03}$
$H_2CO_3 \rightleftharpoons H^+ + HCO_3^-$

$K = \dfrac{a_{H^+} \cdot a_{HCO_3^-}}{a_{H_2CO_3}} = 10^{-6.4}$

Therefore $a_{H^+}.a_{HCO_3^-} = a_{H_2CO_3} \times K$
$\qquad\qquad\qquad\qquad = 10^{-5.03} \times 10^{-6.4}$
$\qquad\qquad\qquad\qquad = 10^{-11.43}$
$a_{H^+} = a_{HCO_3^-} = 10^{-5.7}$
Therefore pH $= 5.7$
CO_2 is less soluble at higher temperature (Box 4.2), but more
soluble at higher pressure (higher P_{CO_2} – Equation 4.15).

Chapter 5

1 2p, 3s, 4f, 5d

2 $_6C$ $1s^2 2s^2 2p^2$
  $_{11}Na$ $1s^2 2s^2 2p^6 3s^1$ or $[Ne]3s^1$
  $_{13}Al$ $[Ne]3s^2 3p^1$
  $_{17}Cl$ $[Ne]3s^2 3p^5$
  $_{18}Ar$ $[Ne]3s^2 3p^6 = [Ar]$
  $_{26}Fe$ $[Ar]4s^2 3d^6$

Chapter 6

1 The Pauli principle dictates that each orbital can hold no more than two electrons:

Z	Name	Core	Valence	Block	Group	Valency
3	Li	$1s^2$	$2s^2$	s	I	1
5	B	$1s^2$	$2s^22p^1$	p	III	3
8	O	$1s^2$	$2s^22p^4$	p	VI	-2
9	F	$1s^2$	$2s^22p^5$	p	VII	-1
14	Si	$1s^22s^22p^6$	$3s^23p^2$	p	IV	4

2 Ti [Ar] $4s^23d^2$: block d

Ni [Ar] $4s^23d^8$: block d

As [Ar] $4s^23d^{10}4p^3$: block p

U [Rn] $7s^26d^15f^3$: block f

3 Na valency = 1. Oxygen = -2, therefore there are 2 atoms of sodium per atom of oxygen:

$$Na_2O \quad x = 2 \quad y = 1$$

Similarly
$$SiO_2 \quad x = 1 \quad y = 2$$
$$SiF_4 \quad x = 1 \quad y = 4$$
$$MgCl_2 \quad x = 1 \quad y = 2$$

Valency of scandium is 3, valency of oxygen = -2. Each Sc combines with 3/2 atoms of oxygen, or 2 Sc atoms combine with 3 oxygen atoms:

$$Sc_2O_3 \quad x = 2 \quad y = 3$$

Similarly
$$P_2O_5 \quad x = 2 \quad y = 5$$
$$BN \quad x = 1 \quad y = 1$$

4 (a) Rewrite Moseley's Law in the linear form:
$$(1/\lambda)^{\frac{1}{2}} = k^{\frac{1}{2}}Z - k^{\frac{1}{2}}\sigma$$

i.e. $y = mx + c$

For each point calculate $(1/\lambda)^{\frac{1}{2}}$ and plot against Z. Points fall on a straight line if the equation holds.

$$k^{\frac{1}{2}} = \text{slope} = \frac{(1/0.561)^{\frac{1}{2}} - (1/0.830)^{\frac{1}{2}}}{47 - 39} = 0.0296 \text{ Å}^{-\frac{1}{2}}$$

$$\sigma = Z - 1/k^{\frac{1}{2}}(1/\lambda)^{\frac{1}{2}}$$

For yttrium (Y) $\sigma = 39 - \dfrac{1.098}{0.0296} = 1.91$

(b) Read off graph at $Z = 43$:
$$(1/\lambda)^{-\frac{1}{2}} = 1.216\,\text{Å}$$

Therefore $\lambda = 0.676\,\text{Å} = 0.676 \times 10^{-10}\,\text{m}$

$$
\begin{aligned}
E_q &= h\nu = hc/\lambda \\
&= \frac{(4.135 \times 10^{-15})(2.997 \times 10^{8})}{0.676 \times 10^{-10}}\,\frac{\text{eV}\,\text{s}\,\text{m}\,\text{s}^{-1}}{\text{m}} \\
&= 18\,332\,\text{eV} = 18.332\,\text{keV}
\end{aligned}
$$

Chapter 7

1 Ionic radii in Box 7.2 lead to following radius ratios:
Si^+: 0.26. Al^{3+}: 0.36 (tet)/0.46 (oct). Ti^{4+}: 0.52. Fe^{3+}: 0.55.
Fe^{2+}: 0.65. Mg^{2+}: 0.61. Ca^{2+}: 0.91. Na^+: 0.94 (8-fold). K^+: 1.20.
Co-ordination numbers: 4: Si^{4+}, Al^{3+}. 6: Al^{3+}, Ti^{4+}, Fe^{3+}, Fe^{2+},
Mg^{2+}. 8: Ca^{2+}, Na^+. 12: K^+.

2 In the higher oxidation state (Fe^{3+}, Eu^{3+}), the ion has fewer
valence electrons and there is less mutual repulsion between them,
so ions are smaller than 2+ ions. Ionic potentials: Fe^{2+}: 23. Fe^{3+}:
41. Eu^{2+}: 16. Eu^{3+}: 28. The 3+ ion has a higher polarizing power
than the 2+ ion and therefore forms a more covalent bond.

3 He–He: van der Waals interaction: very weak, so He is a monatomic
gas at room temperature.
Ho–Ho: electronegativity of Ho is between 1.1 and 1.3, therefore
metallic bond. Crystalline metal at room temperature.
Ge–Ge: metallic/covalent bond. Electronegativity similar to Si in
Figure 7.8b. Semiconductor at room temperature.

4 Using Figures 6.3 and 7.8a:

KCl:	Electroneg. difference = 2.4. 80% ionic.
TiO_2:	Electroneg. difference = 1.9. 65% ionic.
MoS_2:	Electroneg. difference = 0.4.
	Mean electroneg. = 2.4. Submetallic bond.
NiAs:	Electroneg. difference = 0.3.
	Mean electroneg. = 2.1. Submetallic bond.
$CaSO_4$:	(a) S–O bond: <20% ionic. Largely covalent.
	(b) Ca^{2+}–$(SO_4)^{2-}$: More ionic.
$CaSO_4.2H_2O$:	(a) Ca^{2+}–H_2O: Ion–dipole interaction (hydration).
	(b) S–O: As above. (c) $(Ca^{2+}.2H_2O)$–$(SO_4)^{2-}$: Ionic.

Chapter 8

1 $CaFeSi_2O_6$ | Single-chain silicate or ring silicate (in fact, both are
   $NaFeSi_2O_6$ | varieties of pyroxene).
   Edenite: $(AlSi_7O_{22}) \equiv Z_8O_{22}$ — amphibole (double chain).
   Paragonite: $(AlSi_3O_{10}) \equiv Z_4O_{10}$ — sheet silicate (paragonite is the
   Na analogue of muscovite).
   Leucite: $(AlSi_2O_6) \equiv Z_3O_6$ — framework silicate.
   Acmite: $(Si_2O_6) \equiv Z_2O_6$ — chain or ring silicate (acmite is a
   pyroxene).

2 Garnet

Si	17.99
Al	9.57
Ti	0.33
Fe(3)	3.97
Fe(2)	2.92
Mn	0.50
Mg	0.46
Ca	22.58
O	41.27
	99.60

3

	Garnet		Epidote		Pyroxene		Feldspar	
Si	2.9909	} 2.991	3.0035	} 3.004	1.8589	} 2.000	9.5754	
Al^{IV}	–		–		0.1411		6.3618	
								} 16.053
Al^{VI}	1.6515		2.8055		0.0217		–	
Ti	0.0321	} 2.014	–	} 3.044	0.0355		–	
Fe(III)	0.3307		0.2389		0.0272		0.1157	
Fe(II)	0.2436		0.0266		0.6730	} 2.058	–	
Mn	0.0419		0.0005		0.0379		–	
Mg	0.0876	} 2.997	0.0014	} 1.949	0.4312		–	
Ca	2.6235		1.9205		0.7366		2.3799	
Na	–		–		0.0952		1.4753	} 3.886
K	–		–		–		0.0306	
H	–		0.9548	0.955	–		–	

4 Recalculate the three components to 100%:

	per 6 oxygens	100/1.9059
Ca	0.7366	38.65
Mg	0.4312	22.62
$Fe^{2+} + Fe^{3+} + Mn$	0.7381	38.72
Total	1.9059	100.00

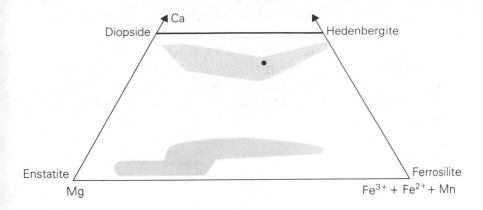

The solid line encloses the *pyroxene quadrilateral*, where most pyroxene analyses would plot. (The stippled areas show the range of pyroxene compositions found in gabbros. The two bands lie on either side of a solvus.)

Chapter 9

1 (a) $Fe_2(SO_4)_3 + FeS_2 \rightarrow 3FeSO_4 + 2S$
   (b) $CuFeS_2 + 2Fe_2(SO_4)_3 \rightarrow CuSO_4 + 5FeSO_4 + 2S$
   (c) $12FeSO_4 + 3O_2 + 6H_2O \rightarrow 4Fe_2(SO_4)_3 + 4Fe(OH)_3$
   (d) $MnO_2 + 4H^+ + 2Fe^{2+} \rightarrow Mn^{2+} + 2H_2O + 2Fe^{3+}$
   (e) $ZnS + 4Fe_2(SO_4)_3 + 4H_2O \rightarrow ZnSO_4 + 8FeSO_4 + 4H_2SO_4$

Chapter 10

1 There are no naturally occurring nuclides (only short-lived radio-nuclides) at N values 19, 35, 39, 45, 61, 89 and 123, nor at Z values 43 (technetium) and 61 (promethium). These are all odd numbers. Odd values of Z and N signify half-filled nuclear orbitals, and such nuclei are more prone to transmutation into other elements by fusion, neutron capture or radioactive decay than even-valued nuclides.

2 (a) The integrated rate equation for radioactive decay (Box 4.2) is:
$$\ln(X_{26}^0/X_{26}) = t\lambda$$
where X_{26} represents the changing concentration of ^{26}Al in a planetary body, and X_{26}^0 is the initial value.
After one half-life (0.7 million years):
$$\ln(1/0.5) = 0.7 \times 10^6 \lambda$$
Therefore $\lambda = 0.99 \times 10^{-6}$ year^{-1}
(b) Heat output is proportional to the rate of decay of ^{26}Al and

therefore to its abundance. When ^{26}Al has decayed to 1% of original value:

$$\ln(1/0.01) = 0.99 \times 10^{-6} t$$

Therefore $t = \ln(100) \times 1.01 \times 10^6$ years $= 4.6$ million years. For this heat source to have been significant during planet formation, planets must have formed within a few million years of nucleosynthesis of ^{26}Al.

Appendix A
Glossary

α-emitter A radioactive nuclide which decays by emitting α-particles. See Box 10.1.

α-particle (alpha particle, α) See Box 10.1.

absolute Measured from a fundamental zero-point, e.g. 'absolute temperature'.

accessory mineral Mineral occurring in small amounts in a rock, not relevant to its identification.

acid, acidic See Appendix C.

adiabatic Describes a process in which the system of interest exchanges no heat with its surroundings.

aerosol A colloidal suspension of fine droplets or particles dispersed in a gas (usually air). See Box 4.3.

alkali, alkaline See Appendix C.

amorphous (of a solid) Lacking crystalline structure and morphology.

amphoteric See Appendix C.

anaerobic In the absence of oxygen (= anoxic).

anhydrous Incorporating no water or OH^-. See **dehydration**.

anion A negatively charged ion, produced when a neutral atom accepts one or more additional electrons.

anisotropic (of a material, e.g. a crystal) Having physical properties that vary according to the direction of measurement.

atomic number (Z) The number of protons in the nucleus of an atom. Also the total number of electrons in the neutral atom.

atomic weight See **relative atomic mass**.

β-particle (beta-particle, β^-) A high-energy electron released by a nuclear reaction.

bar A unit of pressure widely used in petrological literature. It approximates to one atmosphere. It has been superseded in the SI system by the pascal $(Pa) = N\,m^{-2}$. $1\ bar = 10^5\,Pa$.

base, basic See Appendix C.

birefringence (a) A property of a crystal whereby the refractive index varies according to the vibration direction of the incident polarized light. (b)

The numerical difference between the maximum and minimum refractive indices.

catalyst A substance that accelerates a chemical reaction without being consumed by it.

cation A positively charged ion, produced by the loss of one or more electrons from a neutral atom.

CFC Chlorofluorocarbon. Class of inert organic compounds used as refrigerants, foam expanders and aerosol propellants.

component The chemical components of a **system** comprise the minimum number of chemical (atomic, isotopic or molecular) species required to specify completely the compositions of all the **phases** present.

compound A substance in which different elements are combined in specific proportions.

concentration A parameter indicating how much of a particular chemical species (component) is present in a unit amount of the medium (phase) in which it resides (which may be a gas, a liquid or a solid). See Chapter 4.

condensed phases Liquid and solid phases, in which the atoms or molecules are in mutual contact. See Box 1.2.

connate (of waters) Trapped in the interstices of a sedimentary rock at the time of deposition.

co-ordination polyhedron The hypothetical three-dimensional shape obtained by joining the centres of the anions immediately surrounding a cation (or vice versa).

cryptocrystalline Consisting of crystals too small to be readily seen under the optical microscope.

dehydration The removal of water from a substance. In mineralogy, a dehydration reaction is one in which a hydrous (water-bearing) mineral such as mica breaks down at high temperature to form an assemblage of anhydrous (water-free) minerals plus water vapour. (See Figure 2.3)

derivative An expression giving the rate of change of one variable in relation to another upon which it depends. Velocity, the rate at which position x changes with respect to time t, can be regarded as the derivative of x with respect to t $\left(\text{written } \dfrac{dx}{dt}\right)$. See Appendix B and D; Waltham (1994) *Mathematics: a simple tool for geologists*, Chapman & Hall, London.

diatomic (of molecules of gas) Consisting of two atoms.

differential equation An equation in which one or more terms is a **derivative**.

differentiation (a) Mathematical. The process of calculating the gradient or derivative of a dependent variable with respect to another variable. (b) Geochemical. The physical segregation of one type of material from another. A homogeneous or intimately mixed body becomes segregated into two bodies or phases of differing composition. Hence **differentiated**.

diffusion The dispersal of one substance in another by atom-by-atom (or molecule-by-molecule) migration.

dimer A **polymer** consisting of only two basic units.

dipole A system of two electrostatic charges of equal but opposite magnitude, held a specific distance apart. Molecules may be permanent dipoles (owing to internal differences in electronegativity between atoms, as in water) or induced dipoles, caused by the **polarizing** effect of an electrostatic field.

dissociation   A reaction in which a compound decomposes into two or more simpler species. See Chapter 4.

divalent   (of an atom or element) Having a **valency** of 2.

dm^{-3}   'per cubic decimetre' = 'per litre'. See Appendix B.

electrolysis   The extraction of an element (e.g. Cl) from a solution containing its ions (Cl$^-$) by passage of an electric current.

electrolyte   A compound which (as solid, melt or solution) conducts electricity. See Appendix C.

electron (e$^-$ or β^-)   A negatively charged atomic particle responsible for bonding between atoms. See Table 5.1.

electron core   The tightly held inner electron shells not participating in bonding.

electron volt (eV)   The unit of energy used to measure quantum energies and the energy differences within atoms. The kinetic energy possessed by an electron that has been accelerated from rest by an electrostatic field of one volt: $1\,\mathrm{eV} = 1.6021 \times 10^{-19}\,\mathrm{J}$.

element   A substance consisting of a single type of atom (all atoms having the same atomic number Z).

empirical   Determined by experiment or observation, not by theoretical reasoning or calculation.

end-member   One of two or more chemical **components** or formulae in terms of which the composition of a solid solution may be expressed.

endothermic   (of a reaction) Absorbing heat (ΔH positive).

energy   See Chapter 1.

enthalpy (H)   See Chapter 1.

entropy (S)   The internal disorder of a substance (Box 1.2).

eutectic   An invariant point in a phase diagram where the melt field projects to the lowest temperature (e.g. E in Figure 2.4). A eutectic marks the final destination of a melt undergoing fractional crystallization, or the initial melt composition during partial melting (Box 2.4).

evaporite   A sedimentary rock consisting of mineral salts deposited from supersaturated brine as a result of evaporation.

exothermic   (of a reaction) Giving out heat (ΔH negative).

exsolution   (-lamellae) The unmixing of a homogeneous phase into two immiscible phases, commonly as a lamellar intergrowth (e.g. perthite).

extensive   (of variables) Depending in magnitude on the size of the system being considered (e.g. mass).

ferromagnesian   (of silicate minerals) Containing essential Mg and Fe: olivine, pyroxene, amphibole and dark mica.

fluid   Either liquid or gas. The term is applied to anything that can flow. In geology, it often connotes a supercritical aqueous fluid (see Box 2.2).

fluorescence   The excitation of light emission by absorption of shorter-wavelength light.

fractionation   The progressive separation of one type of element from another type. (Hence 'fractionated', the antonym of **primitive**.)

free energy (G)   See Chapter 1.

geothermometry   Measuring the temperature of formation of a rock (the temperature at which its mineral assemblage equilibrated).

glass   Material having the disordered structure of a melt but the mechanical properties of a solid.

hydration   Incorporating water or OH^-. See **dehydration**. The stabilization of an ion in aqueous solution by the gathering of polar water molecules around it. See Box 4.1.

hydrothermal   Refers to hot saline aqueous fluids circulating in the crust, or to veins and ore deposits crystallized from such fluids.

hydroxyl   The OH group or OH^- ion.

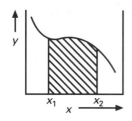

integral   The graph shows y varying as a function of x. The integral of y is an algebraic function expressing the area beneath the y curve between any specified limits x_1 and x_2. It is written $\int_{x_1}^{x_2} y\,dx$.

integration   The mathematical process of calculating the **integral** of, or the area under, a curve. Integration can be considered the reverse of differentiation: thus the integral of velocity as a function of time indicates the distance travelled.

intensive   (of a variable) Having a magnitude independent of the size of the system considered (e.g. temperature).

interstitial   Situated in cavities between larger atoms or ions in a regular crystal structure.

inversion   Recrystallization of a crystalline phase to a more stable polymorph.

ion   Atom with a net electrical charge, due to the acquisition or loss of electron(s). See **anion** and **cation**.

ionic potential   Ratio of an ion's nominal charge over its ionic radius.

ionization energy   The energy (in eV or $kJ\,mol^{-1}$) required to remove the most easily detached electron from an atom to a state of rest at infinity (Figure 5.6).

ionization potential   Obsolete synonym for **ionization energy**.

iso-   Prefix meaning 'the same . . .'.

isobar, isobaric   (a) A pressure contour, a hypothetical line or surface in $P-T-X$ space on which pressure is everywhere uniform. 'Isobaric' signifies a process which occurs at constant pressure. (b) Nuclide having the same mass number A as another.

isotherm, isothermal   A temperature contour in a phase diagram. A line or surface on which temperature is constant. 'Isothermal' signifies a process operating (or a phase diagram constructed) at constant temperature.

isotone   See Box 10.1.

isotope   Isotopes are subsets of an element whose nuclei differ in neutron number. See Box 10.1.

isotopic tracer   An isotopic ratio differing between one type of source material and another, that indicates the derivation of a rock or solution. See Box 9.7.

kilobar (kbar, kb) $1\,kb = 10^8\,Pa$. See **bar**.

kilo-electron volt (keV)   1000 electron volts.

kinetic energy (E_k)   The energy a body possesses by virtue of its motion. See Chapter 1.

lanthanide   One of fourteen elements following lanthanum in the Periodic Table, characterized by the entry of electrons into the 4f orbitals.

latent heat of vaporization   The heat required to transform 1 kg of a liquid into a gas at constant temperature ($J\,kg^{-1}$).

ligand   (in a co-ordination complex or crystal) The species surrounding the central ion or atom. For example in the dissolved complex $Cu(HS)_2^-$, the ligand is HS^-, two of which are attached to the central Cu^+ ion.

linear   A relationship between y and x is said to be linear if plotting y against x yields a straight line.

liquidus   The temperature at which the first crystals begin to form in a cooling melt. A liquidus curve (or surface) shows the variation of liquidus temperature with composition or with pressure.

macroscopic   Visible to the naked eye or measurable with normal laboratory apparatus, as opposed to **microscopic**.

magma   Any sort of igneous melt, including suspended crystals or vapour bubbles.

mass spectrometer   An instrument that separates and analyses the isotopes of an element by accelerating ions into a magnetic field under vacuum. Ions of different mass (A) but the same charge (q) follow trajectories which differ according to the A/q ratio. Output consists of ratios of isotopic abundances.

megascopic   See **macroscopic**.

metalloid   A metal having both metallic and non-metallic properties.

microscopic   Visible only under the microscope. Too small to see or measure with normal apparatus.

miscible   Two compounds are said to be miscible if they can be combined in any desired proportions to form a single, stable, homogeneous phase.

mole   (abbr. 'mol') An amount of a compound (or element) whose mass, expressed in grams, is numerically equal to its **relative molecular (or atomic) mass**.

molecular weight   See **relative molecular mass**.

monatomic   Gases consisting of separate atoms, not combined in molecules.

monovalent   (of an atom or element) Having a **valency** of 1.

native (of an element)   Occurring naturally in the uncombined state (not as a compound).

neutron (n)   Uncharged nuclear particle, marginally heavier than the proton.

neutron number (N)   The number of neutrons in a nucleus.

nucleon   A term encompassing both protons and neutrons in a nucleus.

nuclide   A substance consisting of atoms with particular values of Z and N, i.e. a specific isotope of a specific element.

orbital   See Chapter 5.

organo-metallic   Organic compound incorporating metal atoms.

oxidation state   (of an atom in a molecule/compound) The hypothetical charge the atom would possess if the compound were held together by purely ionic bonds.

oxide   Compound of an element chemically bound to oxygen. See Figure 9.3.

partial pressure (P_i)   The pressure exerted individually by a volatile component i in a gas or other phase.

parts per million (ppm)   Unit of (low) concentration. $1\,ppm = 1\,g$ per $10^6\,g$ $= 1\,\mu g$ per g $= 1\,mg$ per kg. $10^3\,ppm = 0.1\%$.

pH See Appendix C.

phase (chemical usage) A part or parts of a **system** occupying a specific volume and having uniform (or continuously varying, as in a zoned olivine crystal) physical and chemical characteristics which distinguish it (them) from all other parts of the system.

phase boundary Line or surface on a phase diagram marking the limit of $P-T-X$ stability of a particular phase or assemblage.

photon Quantum of light. Behaves in some respects like a particle.

polarizability Susceptibility of a substance to **polarization** on the atomic or molecular scale: a measure of how easily electron density can be deformed in an electric field.

polarization (a) Attraction of negative charge toward one side of an atom, ion or bond, leaving the other side with a net positive charge. (b) A light beam whose electric vector vibrates only in one plane is said to be plane-polarized.

polymer Material whose molecules are built up from a series of identical smaller units. Thus polythene consists of chains of ≤ 80 ethylene (C_2H_4) molecules bonded together.

polymorph One of several alternative crystal structures which a given substance can adopt.

positron (β^+) Anti-matter counterpart of an electron.

precipitate, precipitation Formation of insoluble solid from solution, or formation of liquid droplets from vapour. Requires solution or vapour to be **supersaturated** (Chapter 4).

primitive A material that has undergone little or no **differentiation** or **fractionation**.

prograde (of metamorphic reactions) Proceeding to higher grade, or higher temperature.

proton (p^+) Positively charged nuclear particle.

pseudo-binary, pseudo-ternary System which for practical purposes approximates to a two-component (binary) system, but which in certain details exhibits more complex behaviour e.g. crystallizing phases whose compositions lie outside the system. Likewise 'pseudo-ternary'.

radioactive Undergoing spontaneous nuclear decay.

radiogenic The proportion of a stable **isotope** that is the product of the decay of a **radioactive nuclide**.

radionuclide **Nuclide** with unstable nuclei, undergoing **radioactive** decay.

redox A term covering *red*uction and *ox*idation reactions.

refractory (of elements or their compounds) Having particularly high melting and vaporization temperatures.

relative atomic mass (A_r) Mass of an atom, expressed on a scale in which $^{12}C = 12.0000$.

relative molecular mass, relative molar mass (M_r) Mass of a molecule expressed on scale in which $^{12}C = 12.0000$.

salt See Appendix C.

saturated Solution containing the maximum solute content. It coexists stably with excess solute as solid or gas.

screening Lessening of the electrostatic interaction between two charges owing to another charge interposed between them. Outer electrons in an atom are screened from the nucleus by inner electron shells, making them easier to detach.

secular (of variations in a physical quantity) Non-periodic, changing in the same direction over a period of time.

sialic Rich in Si and Al (e.g. continental crust).

silica Silicon dioxide (SiO_2), in any crystalline or **amorphous** form.

silicate Compound in which one (or more than one) metal is combined with silicon and oxygen. See Chapter 8.

solidus Temperature at which the last fraction of melt crystallizes, or the temperature at which a substance begins to melt. (Cf. **liquidus**).

solute The dissolved species in a solution.

solvation The formation of a sheath of polar solvent molecules around a dissolved ion, inhibiting reaction with other ions and stabilizing the ion in solution. See **hydration** and Box 4.1.

solvent The dominant component of a solution, the medium in which dissolved species are dispersed.

solvus A line or surface in a phase diagram indicating the compositions of immiscible phases in mutual equilibrium at each temperature.

specific heat The heat required to raise the temperature of 1 kg of a substance by one kelvin ($J\,kg^{-1}\,K^{-1}$).

spectrum The various wavelength components present in a beam of light or other electromagnetic radiation (such as an X-ray beam), displayed in order of their wavelength or photon energy. The emission spectrum of an element consists of a series of lines or peaks, representing the specific wavelengths (or photon energies) that are characteristic of that element (e.g. Figure 6.5).

stoichiometry, stoichiometric The proportions, determined by valency, in which elements combine in a compound.

sublimate Solid that crystallizes directly from the vapour, e.g. frost. Sublimation is the vaporization of a solid without intermediate formation of a melt.

supercritical Used of a fluid whose temperature exceeds its critical point. See Box 2.2.

supersaturated (of a solution or solute) Having a molality (or activity) product exceeding the solubility product of that solute at that temperature. Metastable with respect to **precipitation**, tending to precipitate solute until **saturation** is attained.

system Any part of the world to which we wish to confine attention. May refer to a specific domain of composition space, e.g. 'the system $NaAlSi_3O_8$–$CaAl_2Si_2O_8$'.

texture Describes the geometrical relationships between the constituent mineral grains in a rock.

thermal energy Total **kinetic energy** possessed by a substance by virtue of individual molecular motions.

tie-line Isothermal line in a phase diagram linking two phases of different chemical compositions that are in mutual chemical equilibrium at the temperature concerned. See Box 2.3.

trivalent (of atom, element) Having a **valency** of 3.

valence electron An electron occupying an orbital in the valence (highest energy) shell, available for bonding.

valency The number of chemical bonds that an atom (element) can utilize in forming a molecule (compound). Some elements have more than one valency.

zoned, zoning Continuous or abrupt changes in composition within a single phase.

zwitterion An ion carrying both a positive and negative charge (a property of amino acids, Chapter 9).

Appendix B
Mathematics revision

B.1   SI Units of measurement

About 25 years ago, the Système International d'Unités ('SI units') was introduced to promote standardization between sciences. The following are the most important features:

(a)   The basic units are

Dimension	Unit	Symbol
length	metre	m
mass	kilogram	kg
time	second	s
electric current	ampere	A
temperature	kelvin	K

The kelvin is the unit of temperature measured from absolute zero. $T(K) = T(°C) + 273.15$.

Moles are abbreviated to 'mol' in the SI notation.

Volume is expressed in m^3; in solution chemistry, the litre (l) has been superseded by dm^3 (cubic decimetres) in SI notation: $1 dm^3 = 10^{-3} m^3 = 1$ litre.

(b)   Names are given to various 'derived' SI units consisting of combinations of the basic units above. The derived units are inconvenient to remember at first, but make the system much more concise to use. Examples are:

force	newton	N	$kg\,m\,s^{-2} = J\,m^{-1}$
energy	joule	J	$kg\,m^2\,s^{-2} = N\,m$
power	watt	W	$kg\,m^2\,s^{-3} = J\,s^{-1}$
pressure	pascal	Pa	$kg\,m^{-1}\,s^{-2} = N\,m^{-2} = J\,m^{-3}$
electrical potential difference	volt	V	$kg\,m^2\,s^{-3}\,A^{-1} = W\,A^{-1}$

Notice that some derived units have several equivalent forms and may be expressed in alternative ways, either in the basic units or in terms of other derived units. An example is the constant g, the acceleration due to gravity, which may be expressed as $m\,s^{-2}$ or as $N\,kg^{-1}$, the two forms being exactly equivalent.

(c) The units of heat are the same as those of mechanical energy and work, avoiding the need for the 'mechanical equivalent of heat' constant (converting calories to joules) required by earlier systems of units.

(d) The system recognizes prefixes which multiply or divide units by factors of 10^3:

teragram	Tg	$= 10^{12}\,g = 10^9\,kg$
gigametre	Gm	$= 10^9\,m$
megawatt	MW	$= 10^6\,W$
kilojoule	kJ	$= 10^3\,J$
millivolt	mV	$= 10^{-3}\,V$
micrometre	μm	$= 10^{-6}\,m$
nanogram	ng	$= 10^{-9}\,g = 10^{-12}\,kg$
picometre	pm	$= 10^{-12}\,m$

Not all disciplines have adopted the SI system wholeheartedly. Some petrologists, for example, continue to use such units as bars ($10^5\,Pa$) for pressure.

B.2   Gradient of a curve

Many problems in geochemistry require us to determine the slope or **gradient** of a graph. The gradient tells us how rapidly one variable (the dependent variable, usually plotted on the y axis) changes in response to variation in the value of another (the independent variable, plotted on the x axis). If we plot the position y of a car travelling along a straight road against time t, for instance, we get a curve whose gradient at each point along the curve tells us the car's velocity at the instant concerned: the steeper the graph, the faster the car was travelling. The horizontal portions of the curve, on the other hand, indicate when the car was stationary.

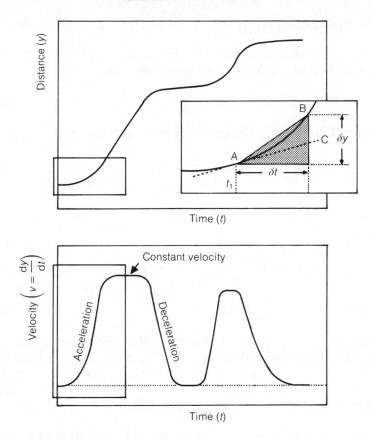

How do we measure the gradient (the car's speed) at a specific point on this curve? The figure shows a portion of the curve enlarged, and we want to measure the gradient at the point t_1. We can begin by determining the slope of the chord AB, as a first approximation to that of the curve:

$$\text{Gradient of chord AB} = \frac{\delta y}{\delta t} \qquad \text{(B.2.1)}$$

δt is an **increment** (a small increase) in t, and δy is the consequent increment in y. If we make δt (and consequently δy) smaller and smaller, the slope of the chord grows closer and closer to the gradient of the curve at t_1. If we reduce δt to the point where it is effectively zero (written symbolically $\delta t \to 0$), the chord coincides with the tangent, AC, which by definition has the same gradient as the curve itself at A. Expressed in terms of such infinitesimal increments, written dy and dt, the slope of the curve at point A is:

$$\text{Gradient at A} = \frac{dy}{dt} = \text{the \textbf{derivative} of } y \text{ with respect to } t. \qquad \text{(B.2.2)}$$

B.2.1 Differentiation

This symbolism takes on more meaning when the variation of y with t can be expressed in the form of an equation. Suppose that

$$y = at^2 + c \qquad (B.2.3)$$

Since point B also lies on the curve, it must also satisfy this equation:

$$(y + \delta y) = a(t + \delta t)^2 + c \qquad (B.2.4)$$
$$= [at^2 + 2at\delta t + a(\delta t)^2] + c$$

Therefore $\qquad \delta y = 2at\delta t + a(\delta t)^2$

so $\qquad \dfrac{\delta y}{\delta t} = 2at + a\delta t$

As $\delta t \to 0$, $\qquad \dfrac{dy}{dt} = 2at \qquad (B.2.5)$

This indicates that the gradient at any specific point t_1 on the curve can be calculated simply by evaluating $2at_1$. The expression $2at$ is called the derivative of $y = at^2 + c$ with respect to t. The algebraic or arithmetical process by which we calcuate the derivative is called **differentiation**, the basic operation in a branch of mathematics called differential calculus.

By similar reasoning one can show that the derivative of $y = at^3$ is $3at^2$, and the derivative of $y = at^4$ is $4at^3$ and so on. In general,

if $\qquad y = bx^n$

then $\qquad \dfrac{dy}{dx} = nbx^{n-1}.$

For example, if $\quad y = ax^3 + bx^2 + cx + d$
$$= 15.2x^3 + 2.9x^2 - 16.8x - 4.3$$

then $\qquad \dfrac{dy}{dx} = 3ax^2 + 2bx + c = 45.6\,x^2 + 5.8x - 16.8$

To determine the numerical value of the gradient a specific value of x, one simply introduces that value of x into the dy/dx equation:

For example $\qquad x = 2.0$

$$\dfrac{dy}{dx} = 182.4 + 11.6 - 16.8 = 177.2.$$

Thus one can evaluate the slope of the curve at any desired point.

For a more extensive introduction to differential calculus the reader should refer to the excellent book by Waltham (1994).

B.2.2 Second derivative

If we differentiate Equation B.2.3, we get a new equation – the derivative – indicating the velocity of the car as a function of time:

$$v = \frac{dy}{dt} = 2at \qquad\qquad (B.2.6)$$

and if we wished to we could plot this as a graph against time (see lower graph). This graph has a uniform gradient and passes through the origin. We can evaluate the gradient by differentiating Equation B.2.6:

$$\frac{dv}{dt} = 2a$$

This 'derivative of a derivative' is known as the **second derivative** of equation B.2.3, and is sometimes written $\frac{d^2y}{dt^2}$. Knowing the value of a, it would be a simple matter to evaluate the gradient.

What physical significance does this second derivative have? The rate at which velocity changes with time is familiar as the **acceleration**. In this particular example, the vehicle is undergoing constant acceleration, so that the velocity is increasing in a linear fashion with time.

The second derivative is an important concept in certain classes of **differential equation** (Box 3.5).

B.2.3 Dimensions and units in calculations

In writing the answer to any numerical problem, one should give two items: the numerical answer and its units. The number by itself is incomplete (unless it is a 'dimensionless number').

One must pay attention to units at every stage in calculation, checking that all the quantities are expressed in compatible units. If one variable is entered in millimetres when it should appear in metres, you are immediately introducing an error of 1000 times.

The procedure for carrying units through a calculation can be illustrated by the Clapeyron equation (Chapter 2). Suppose we wish to know the slope of the phase boundary representing the reaction between two (isochemical) minerals A and B:

Reaction	A $\rightarrow$ B	Units in published tables of S and V
Molar entropies:	S_A S_B	$J\,K^{-1}\,mol^{-1}$
Molar volumes:	V_A V_B	$m^3\,mol^{-1}$

$$\Delta S = S_B - S_A \qquad J\,K^{-1}\,mol^{-1}$$
$$\Delta V = V_B - V_A \qquad m^3\,mol^{-1}$$

Clapeyron equation:

$$\frac{dP}{dT} = \frac{\Delta S}{\Delta V} \qquad \frac{J\,K^{-1}\,mol^{-1}}{m^3\,mol^{-1}}$$

'mol^{-1}' appears on top and bottom, and therefore cancels out, leaving the units for the gradient as $J\,m^{-3}\,K^{-1} = N\,m^{-2}\,K^{-1} = Pa\,K^{-1}$. Using volumes expressed in $cm^{-3}\,mol^{-1}$ would make it necessary to introduce a correction factor of 10^6.

Some physical parameters are 'dimensionless' and therefore have no units. They are pure numbers. Specific gravity (density of a substance/density of pure water at $4\,°C$) is an example. The numerical values of such numbers are independent of the units being used in their computation. Thus in calculating specific gravity, the units of density cancel out, provided that the two densities are expressed *in the same units*.

Graphs must show the units in which each of the variables is expressed. The present practice is to label each axis 'quantity/units', e.g. $T/°C$.

B.2.4   Experimental verification of a theoretical relationship

When a mathematical equation is proposed (commonly on theoretical grounds) to describe a phenomenon, one often wishes to test it against available experimental observations. Does it describe the experimental results accurately, or would some other form of equation be more appropriate? The simplest way to answer this question is to plot the experimental data and the theoretical equation together in a suitable graph.

It is important that this be done in such a way that the form of the graph is linear. To see why, consider the verification of the Arrhenius equation relating rate constant to absolute temperature:

$$k = A\exp\left(-\frac{E_a}{RT}\right) \quad \text{or} \quad k = Ae^{-E_a/RT}$$

If we were to plot experimental results for k against T, the results would lie on a curve:

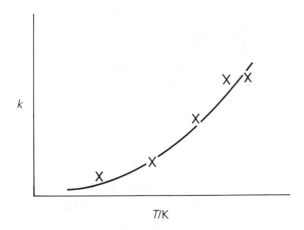

Unless we happened to know the constants A and E_a in advance – and in general we don't – it would be difficult to determine whether the curve defined by the experimental data has the shape predicted by the equation. It would be necessary to use a complicated curve-fitting calculation to establish the agreement between experimental data and the theoretical equation.

Consider an alternative plot. The Arrhenius equation may be written in the log form:

$$\ln k = \ln A + \ln\left[\exp\left(-\frac{E_a}{RT}\right)\right]$$

$$= \ln A - \frac{E_a}{RT}$$

$$(y = c + mx)$$

It is clear that when $\ln k$ is plotted against $1/T$, the form of this equation predicts a straight line, with slope $= -\left(\frac{E_a}{R}\right)$ and intercept $= \ln A$.

Thus there are two reasons for manipulating a theoretical equation into linear form before attempting to verify it against experimental results:

(a) It is easy to see how straight a straight-line graph is. It is much harder to judge the curvature that a curved relationship should have.

(b) It is a simple calculation to determine the values of the constants

281

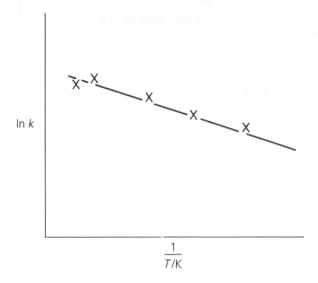

in the equation (e.g. A, E_a) from a straight-line graph. (A linear relation between y and c may be written $y = mx + c$, where m is the gradient and c is the y-axis intercept.) It is harder to extract these constants from a curved graph.

TERNARY DIAGRAMS

Three variables can be plotted in a two-dimensional graph if they add up to 100%. Any composition in a three-component system can therefore be represented in two dimensions, usually in the form of a **ternary diagram** plotted on special triangular graph paper (available commercially). The user labels each apex with one of the components, as shown opposite.

Each apex thus represents 100% of the component with which it is labelled (the Di corner represents 100% diopside). The side opposite represents compositions devoid of the component concerned (in this case, Ab–An mixtures). Lines parallel to this edge are contours representing different Di percentages from 0 to 100%.

To plot a composition such as Di 72%, Ab 19%, An 9% (notice the three co-ordinates must add up to 100%), rule a line horizontally across the diagram at the position equivalent to 72% Di. Rule another line, parallel to the Di–An edge, at the position corresponding to 18% Ab. The intersection with the first line marks the composition being plotted. Note that only two readings need to be plotted; the third, being the difference from 100, is not an independent variable – but it is useful to read it off the diagram (An = 9%) to check that the point has been plotted accurately.

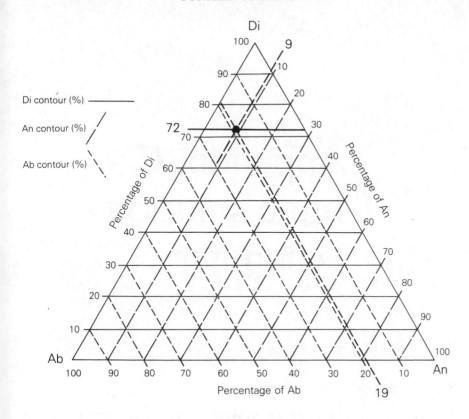

B.3 Further reading

Waltham, D. (1994) *Mathematics: a simple tool for geologists*, Chapman & Hall, London.

Appendix C
Simple solution chemistry

C.1 Acids and bases

C.1.1 Acids

The simplest definition of an **acid** is 'a substance capable of contributing hydrogen ions to a solution or reaction' – that is, an H^+ donor. For example, hydrochloric acid (HCl) when dissolved in water undergoes ionization:

$$HCl \rightarrow H^+ + Cl^-$$

(In fact this is a simplification. Each ion is surrounded in solution by a 'sheath' of water molecules attracted electrostatically by the charge of the ion (**hydration** – Box 4.1). One can represent these hydrated ions as $H^+_{(aq)}$ and $Cl^-_{(aq)}$.

An acid like HCl, which contributes only one hydrogen ion (H^+) per acid molecule to the solution, is said to be **monobasic** (or monoprotic). Sulphuric acid (H_2SO_4) is a **polybasic** (or polyprotic) acid: it may lose both or only one of its hydrogens in reactions:

$$H_2SO_4 + KOH \rightarrow KHSO_4 + H_2O$$
$$\text{potassium}$$
$$\text{bisulphate}$$

$$H_2SO_4 + 2KOH \rightarrow K_2SO_4 + 2H_2O$$
$$\text{potassium}$$
$$\text{sulphate}$$

Potassium bisulphate can contribute further H^+ ions in subsequent reactions: it is itself acidic.

A **base** is a substance that accepts or absorbs H^+ ions, thereby depleting the host solution in free H^+ and making it *less* acidic. The above reaction between KOH and H_2SO_4 may be written:

$$\underset{\text{base}}{(K^+ + OH^-)} + \underset{\text{acid}}{(H^+ + HSO_4^-)} \rightarrow (K^+ + HSO_4^-) + H_2O$$

showing that KOH is behaving as a base, converting H^+ into H_2O.

Other reactions between acids and bases show the same pattern:

$$\underset{\text{base}}{NaOH} + \underset{\text{acid}}{HCl} \rightarrow \underset{\text{salt}}{NaCl} + \underset{\text{water}}{H_2O}$$

When acids and bases react together, they tend to **neutralize** each other, forming **salts** (plus water).

Acids, bases and salts form **electrolyte** solutions, in which the solute is partly or completely ionized, resulting in electrical conductivity. Compounds like HCl which are more or less completely ionized in solution are called 'strong electrolytes'. Weak electrolytes (like carbonic acid, H_2CO_3) are those which exhibit only slight ionization in aqueous solution. Salts are almost always strong electrolytes, but acids and bases may be strong or weak, depending on the bond holding the compound together (Chapter 7).

C.1.2 Ionization of water: pH

Pure water undergoes partial self-ionization:

$$H_2O \rightarrow H^+ + OH^+$$

It is thus a weak electrolyte. The equilibrium constant (Chapter 4) for this reaction is such that the concentrations of free H^+ and OH^- ions in pure water are both about $10^{-7}\ mol\ kg^{-1}$. This can be expressed concisely by saying that the **pH** of pure water is 7, where

$$pH = -\log_{10} m_{H^+}$$

So if a solution has a pH of 2 it means that the concentration of free hydrogen ions (m_{H^+}) is $10^{-2}\ mol\ kg^{-1}$. The pH notation may be used to describe the acidity of a solution. Values of pH less than 7.0 denote higher concentrations of the H^+ ion than are found in pure water (**acidic** behaviour), whereas values above 7.0 indicate lower H^+ concentrations than pure water (**basic** behaviour).

The pH of a solution can be measured in two ways:

pH	Ground waters and precipitation	Body fluids	Foods
	Sea water		
— 9			
— 8	} Limestone areas	Blood	Fresh eggs
7			
— 6	} Unpolluted rain;		Cows' milk
— 5	} most ground waters		Beer
— 4	Peat water		Tomatoes
— 3	Mine waters	} Stomach	Vinegar
— 2	Acid Los Angeles smog	contents	Lemons
— 1			Limes

(a) using a paper treated with a pH-sensitive dye whose colour indicates the pH of the solution (litmus paper is the traditional paper, but more specific pH papers are now available);

(b) using a special meter called a pH meter, which gives a direct reading of pH.

The table above shows the pHs of some familiar solutions.

To summarize, an acid solution is one whose H^+ concentration is greater than that found in pure water ($10^{-7}\,mol\,kg^{-1}$). An acid solute is one that raises the H^+ concentration of a solution, and a base is one that depresses it. An equivalent definition of a base is a solute that increases the concentration of hydroxyl (OH^-) ions in solution: because additional OH^- will associate with H^+ ions to form water, the addition of OH^- is equivalent to a reduction of m_{H^+}.

The term **alkali** describes a water-soluble base, in particular the hydroxides of sodium (Na) and potassium (K). For this reason Na and K are known as **alkali metals**; in geological usage this term is often abbreviated to 'alkalis'. Sodium and potassium form basic oxides (Box 8.1).

C.2   Further reading

For further background, the following book is recommended:
Hill, G. (1986) *Chemistry counts*, Hodder & Stoughton, London.

Appendix D
Alphabetical list of chemical symbols and element names, with atomic number and relative atomic mass

Symbol	Name	Z	Relative atomic mass[†]
Ac	actinium	89*	227.03
Ag	silver	47	107.87
Al	aluminium	13	26.98
Ar	argon	18	39.95[‡]
As	arsenic	33	74.92
At	astatine	85*	210.
Au	gold	79	196.97
B	boron	5	10.81
Ba	barium	56	137.34
Be	beryllium	4	9.01
Bi	bismuth	83	208.98
Br	bromine	35	79.91
C	carbon	6	12.01
Ca	calcium	20	40.08

Symbol	Name	Z	Relative atomic mass
Cd	cadmium	48	112.40
Ce	cerium	58	140.12
Cl	chlorine	17	35.45
Co	cobalt	27	58.93
Cr	chromium	24	52.01
Cs	cesium	55	132.91
Cu	copper	29	63.54
Dy	dysprosium	66	162.50
Er	erbium	68	167.26
Eu	europium	63	151.96
F	fluorine	9	19.00
Fe	iron	26	55.85
Fr	francium	87★	223.
Ga	gallium	31	69.72
Gd	gadolinium	64	157.25
Ge	germanium	32	72.59
H	hydrogen	1	1.008
He	helium	2	4.00‡
Hf	hafnium	72	178.49
Hg	mercury	80	200.59
Ho	holmium	67	164.93
I	iodine	53	126.90
In	indium	49	114.82
Ir	iridium	77	192.2
K	potassium	19	39.10
Kr	krypton	36	83.80
La	lanthanum	57	138.91
Li	lithium	3	6.94
Lu	lutetium	71	174.97
Mg	magnesium	12	24.31
Mn	manganese	25	54.94
Mo	molybdenum	42	95.94
N	nitrogen	7	14.01
Na	sodium	11	22.99

Symbol	Name	Z	Relative atomic mass
Nb	niobium	41	92.91
Nd	neodymium	60	144.24
Ne	neon	10	20.18
Ni	nickel	28	58.71
Np	neptunium	93*	237.05
O	oxygen	8	16.00
Os	osmium	76	190.2
P	phosphorus	15	30.97
Pa	protactinium	91*	231.04
Pb	lead	82	207.19‡
Pd	palladium	46	106.4
Pm	promethium	61*	146.92
Po	polonium	84*	210.
Pr	praseodymium	59	140.91
Pt	platinum	78	195.09
Pu	plutonium	94*	239.05
Ra	radium	88*	226.03
Rb	rubidium	37	85.47
Re	rhenium	75	186.20
Rh	rhodium	45	102.91
Rn	radon	86*	222.
Ru	ruthenium	44	101.07
S	sulphur	16	32.06
Sb	antimony	51	121.75
Sc	scandium	21	44.96
Se	selenium	34	78.96
Si	silicon	14	28.09
Sm	samarium	62	150.35
Sn	tin	50	118.69
Sr	strontium	38	87.62‡
Ta	tantalum	73	180.95
Tb	terbium	65	158.92
Tc	technetium	43*	98.91
Te	tellurium	52	127.60
Th	thorium	90*	232.04
Ti	titanium	22	47.90
Tl	thallium	81	204.37
Tm	thulium	69	168.93

289

Symbol	Name	Z	Relative atomic mass
U	uranium	92*	238.03
V	vanadium	23	50.94
W	tungsten	74	183.85
Xe	xenon	54	131.30
Y	yttrium	39	88.91
Yb	ytterbium	70	173.04
Zn	zinc	30	65.37
Zr	zirconium	40	91.22

* Signifies an element with no stable isotope.
† Relative to $^{12}C = 12.000$.
‡ RAM may vary slightly with proportion of radiogenic isotopes.

Index

Standard-type numbers refer to page numbers, *italic* numbers refer to *figures* and **bold** numbers refer to **boxes**.

INDEX

Periodic Table of the elements

College Library